Alkenes - Recent Advances, New Perspectives and Applications

Edited by Reza Davarnejad

Published in London, United Kingdom

IntechOpen

Supporting open minds since 2005

Alkenes - Recent Advances, New Perspectives and Applications
http://dx.doi.org/10.5772/intechopen.94671
Edited by Reza Davarnejad

Contributors
Bhisma K. Patel, Tipu Alam, Amitava Rakshit, Reza Davarnejad, Jamal Azizi, Shaghayegh Bahari, Zahra Balzade, Farhad Sharif, Seyed Reza Ghaffarian Anbaran, Wangjing Ma, Duanda Wang, Jun Zhao, Bonan Liu, Lu Zhang, Lei Zhang, Shikha Agarwal, Jay Soni, Pankaj Teli, Nusrat Sahiba, Ayushi Sethiya

First published in London, United Kingdom, 2021 by IntechOpen
IntechOpen is the global imprint of INTECHOPEN LIMITED, registered in England and Wales, registration number: 11086078, 5 Princes Gate Court, London, SW7 2QJ, United Kingdom
Printed in Croatia

British Library Cataloguing-in-Publication Data
A catalogue record for this book is available from the British Library

Additional hard and PDF copies can be obtained from orders@intechopen.com

Alkenes - Recent Advances, New Perspectives and Applications
Edited by Reza Davarnejad
p. cm.
Print ISBN 978-1-83969-535-3
Online ISBN 978-1-83969-536-0
eBook (PDF) ISBN 978-1-83969-537-7

For EU product safety concerns:
IN TECH d.o.o., Prolaz Marije Krucifikse Kozulić 3, 51000 Rijeka, Croatia, info@intechopen.com or visit our website at intechopen.com.

Meet the editor

Dr. Reza Davarnejad obtained a Ph.D. in Chemical Engineering from Universiti Sains Malaysia (USM)in 2010 where he focused his studies on supercritical technology, thermodynamics, and biotechnology. He has more than 120 publications and research activities to his credit. Dr. Davarnejad has been an associate professor at Arak University, Iran, since 2016. From 2018 to 2020, he was head of the Chemical Engineering Department at the same university. His research interests include environmental engineering (wastewater treatment) through advanced oxidation processes such as electro-Fenton and adsorption methods. Dr. Davarnejad recently joined the nanobiotechnology group at Arak University as a founding member.

Contents

Preface

This book illustrates new findings on alkenes, including information on the synthesis of alkenes, olefins, and polyolefins.

Chapter 1 presents an overview of olefins and their preparation.

Chapter 2 shows the oxidation of alkenes as an important reaction used to develop epoxides, carbonyls, allylic compounds, 1, 2-diols, and so on. In this work, metal-catalyzed oxidation of alkenes is examined. This chapter highlights the metal-catalyzed green oxidation of alkenes and provides new strategies for the functionalization and transformation of alkenes.

Chapter 3 expresses the photo-redox process via single-electron transfer (SET). It studies the synthesis of targeted organic compounds due to their environmental friendliness and sustainability. According to this chapter, the construction of the C–N bond via visible-light-mediated difunctionalization of alkenes would be useful for medicinal and synthetic organic chemists and inspire further reaction development.

Chapter 4 describes the catalytic isomerization of olefins and their derivatives. According to this chapter, carbon-carbon double bond (CCDB) isomerization is a method for synthesizing new organic compounds from olefins and their derivatives based on C=C migration along the carbon chain and cis/trans transform. In fact, it plays a vital role in the fields of organic synthesis, synthesis of daily chemicals, development of raw oil, and synthesis of natural products. Chapter 5 presents research on extending alkenes' value chain to functionalized polyolefins. The chapter reviews the commercial specialty and functions of polyolefins as well as future prospects.

Finally, Chapter 6 discusses some industrial aspects of olefins' production and the process detail through naphtha feed.

Reza Davarnejad
Chemical Engineering Department,
Arak University,
Arak, Iran

Section 1

Introduction

Chapter 1

Introductory Chapter: Olefins - Past, Now and Future

Reza Davarnejad

1. Introduction

The previous book published by the same publisher illustrates some findings (until that date) on the alkenes molecular structure (synthesis and application) [1]; now, the current book will try to show some new techniques on the alkenes separation and purification although some industrial aspects will also be considered. Therefore, alkenes (olefins) quality, their separation and purification techniques, and economic and environmental aspects of their production will be studied.

2. Various techniques in olefins production

Alkenes (olefins) play an important role in the human life and their consumption is increasing. Most industries produce the olefins through steam cracking (SC) of hydrocarbons, while few of them are investing in alternative technologies and feedstocks. These technologies have been encouraged due to the abundance of cheap propane, ethane, and methane from shale and stranded gases. From the economic point of view, methane is an interesting starting material if products can be produced from it [2]. According to the literature, the future for proven technologies such as Fischer-Tropsch synthesis (FTS) or methanol to gasoline is vague. Therefore, some industries tried to produce light olefins by the catalytic dehydrogenation of propane. Ethylene and propylene are the main olefins, with a production rate of 1.5×10^8 t/yr and 8×10^7 t/yr, respectively [3]. These are expected to increase with the increasing global population combined with rising living standards [4]. In fact, the light olefins and their derivatives are the modern life basis. The olefin production traditionally depends on natural gas processing products or crude oil fractions based on steam cracking technology. In this process, hydrocarbons that primarily originate from fossil resources are cracked at elevated temperatures in tubular reactors suspended in a gas-fired furnace. However, this process has wieldy been applied and optimized, but its economical issue is challengeable. Furthermore, alternative and more sustainable processes and feedstocks would be required to fulfill the future demand for the chemicals. The methane supply has enormously increased since 2008 and its price has considerably dropped [5, 6]. Therefore, shale gas has been bolded as an interesting cost-competitive feedstock for valorizing methane in the form of olefins and higher hydrocarbons. Moreover, the abundance of cheap ethane from the shale gas has enabled cheap low-olefin production through the steam cracking and has had a proper impact on olefin market of some countries (such as the USA) [7–9]. Ethane steam crackers were found to add around 1×10^7 t of ethylene capacity in 2020 in the USA. The steam cracking of hydrocarbons would be the main technique in the light olefin production. Therefore, lighter feedstocks developed on-purpose of higher olefins and even propylene production [7, 10]. Petrochemical producers can skip the refining step and reduce their production costs by using the recent process [11].

Since the use of coal as a feedstock is discouraged from the environmental aspects, a lot of researchers and engineers are trying to explore alternative eco-friend production techniques. However, energy and chemicals production is an important issue for the world's future, but environmental concerns on the various techniques should be noted. According to the literature, each of these alternative olefin production approaches benefits from the abundance of propane, ethane, and methane from shale gas and stranded gas although the relevance of each approach to lighter feedstock utilization in the steam cracking of hydrocarbons results in reducing the production of important coproducts. Ethylene and propylene would mainly be produced by the steam cracking of hydrocarbons. Not only the steam cracking of hydrocarbons is the most important process in the petrochemical industries but also is the most energy-intensive process. In fact, it is a leading technology for light olefin production. In this process, a hydrocarbon feedstock is mixed with steam and cracked at elevated temperatures in a tubular reactor. The feedstock is from light alkanes (such as ethane and propane) to complex mixtures (such as naphtha) and gas oils [12]. The global propylene and crude C_4 production capacity (for the production of ethylene in steam crackers) has an impact by feed shifting from naphtha to light feeds. However, the steam cracking feedstock makes a shift to ethane low price feedstocks, but shipping and transport *via* pipes of ethane also provided several advantages such as local low ethane prices. On the other hand, the crude oil large availability caused producers of ethylene (such as ExxonMobil) to shift to this process, as well [2]. However, some ethylene producers located in India, Brazil, Canada, and several European countries are going to import ethane but some companies (in China) mainly depend on naphtha cracking. This can limit their competitiveness in the world because the other companies [in Iran (Asaluyeh county)] are producing various olefins from natural gas that is a cheap source in some countries having huge natural gas resources.

Finally, several technologies such as the catalytic dehydrogenation of light alkanes, oxidative coupling of methane (OCM), and syngas-based routes such as the Fischer-Tropsch synthesis (FTS) and methanol synthesis followed by methanol to olefins (MTO) have recently been paid attention [13–16]. According to the literature, FTS is not a selective technique for the light olefin production due to producing a considerable amount of the other hydrocarbons, which can be used as fuel. This process is improving with catalyst design. OCM as an alternative process should be modified in terms of the reactor because of the strong exothermicity of its reaction. However, it seems that steam cracking still is a predominant process for the olefin production (with a mainly lighter feedstock) but a lot of researchers are studying alternative technologies to remove some bottlenecks in the process efficiency and CO_2 footprint.

Author details

Reza Davarnejad
Faculty of Engineering, Department of Chemical Engineering, Arak University, Arak, Iran

*Address all correspondence to: davarnejad.usm.ac@gmail.com

References

[1] Davarnejad R, Sajjadi B. Alkenes. 1st ed. Croatia: Intechopen; 2018. pp. 1-170

[2] Amghizar I, Vandewalle LA, Van Geem KM, Marin GB. New trends in olefin production. Engineering. 2017;**3**:171-178

[3] Zimmermann H, Walzl R. Ethylene. In: Ullmann's Encyclopedia of Industrial Chemistry. Weinheim: Wiley-VCH Verlag GmbH and Co. KGaA; 2009

[4] BP Statistical Review of World Energy, BP Technical Report. London: BP Plc; 2013

[5] United States Energy Information Administration. Annual Energy Outlook 2015 with Projections to 2040. Washington DC: United States Energy Information Administration; 2015

[6] Sattler JJ, Ruiz-Martinez J, Santillan-Jimenez E, Weckhuysen BM. Catalytic dehydrogenation of light alkanes on metals and metal oxides. Chemical Reviews. 2014;**114**: 10613-10653

[7] Bruijnincx PC, Weckhuysen BM. Shale gas revolution: An opportunity for the production of biobased chemicals? Angewandte Chemie (International Ed. in English). 2013;**52**:11980-11987

[8] Siirola JJ. The impact of shale gas in the chemical industry. AICHE Journal. 2014;**60**:810-819

[9] Yang CJ. US shale gas versus China's coal as chemical feedstock. Environmental Science & Technology. 2015;**49**:9501-9502

[10] Ding J, Hua W. Game changers of the C3 value chain: Gas, coal, and biotechnologies. Chemical Engineering and Technology. 2013;**36**:83-90

[11] Available from: http://news.ihsmarkit.com/press-release/new-exxonmobil-and-saudi-aramco-technologies-produce-ethylene-directly-crude-oil-cutti

[12] Van Geem KM, Reyniers MF, Marin GB. Challenges of modeling steam cracking of heavy feedstocks. Oil & Gas Science and Technology—Revue IFP Energies Nouvelles. 2008;**63**:79-94

[13] Al-Salem S, Lettieri P, Baeyens J. Recycling and recovery routes of plastic solid waste (PSW): A review. Waste Management. 2009;**29**:2625-2643

[14] Al-Salem S, Lettieri P, Baeyens J. The valorization of plastic solid waste (PSW) by primary to quaternary routes: From re-use to energy and chemicals. Progress in Energy and Combustion Science. 2010;**36**:103-129

[15] Bridgwater AV. Review of fast pyrolysis of biomass and product upgrading. Biomass & Bioenergy. 2012;**38**:68-94

[16] Putro JN, Soetaredjo FE, Lin SY, Ju YH, Ismadji S. Pretreatment and conversion of lignocellulose biomass into valuable chemicals. RSC Advances. 2016;**6**:46834-46852

Section 2

Synthesis

Chapter 2

Metal Catalyzed Oxidation Reactions of Alkenes Using Eco-Friendly Oxidants

Jay Soni, Pankaj Teli, Nusrat Sahiba, Ayushi Sethiya and Shikha Agarwal

Abstract

Oxidation of alkenes is an important reaction in academia, industry and science as it is used to develop epoxides, carbonyls, allylic compounds, 1,2-diols, etc. Metal catalyzed oxidation of alkenes has aroused as a significant tool in modern organic synthesis. Several techniques are available; however some of them suffer from few shortcomings viz. high cost, toxic nature, harsh reaction condition, solid waste generation, etc. In view of these drawbacks, green oxidants i.e. O_2, H_2O_2, TBHP, etc. have shown noteworthy prospects due to their nature, low cost, high atom economy and high sustainability in metal catalyzed reactions. This chapter highlights the metal catalyzed green oxidation of alkenes and shall provide new strategies for the functionalization and transformation of alkenes.

Keywords: Oxidation, alkenes, green oxidants, metal based catalysts, organic synthesis

1. Introduction

The carbon–carbon (C-C) bond is the fundamental unit of valuable organic molecules and possesses significant and diversified applications in nature. Alkenes are the copious portion of organic compounds that are produced abundantly from renewable resources and petrochemical feedstocks [1]. The unique reactivity profile of alkenes is responsible for the diversification and modification in the skeleton that makes them essential in organic synthesis. As a consequence, they have been implemented as starting materials in the synthesis of a wide range of organic chemical building blocks. Various types of molecules such as ethers, halo-ether, alcohol, diol, alkane, and halo compounds have been produced from alkenes via different chemical transformations like addition, substitution, oxidation, polymerization and so on.

Oxidation of alkenes is recognized as a powerful and straightforward tool for the construction of carbon–oxygen (C-O) bonds to develop epoxides, carbonyls, 1,2-diols, and allylic compounds, etc [2–4]. During the past decades, researchers paid much attention on this subject and several methodologies have been proposed to achieve the selective oxidation of alkenes. Several strong oxidizing agents like OsO_4, RuO_4, $NaIO_4$, CrO_2Cl_2 are used in stoichiometric amount for this purpose [5–9]. These protocols suffer from one or more drawbacks like expensive and toxic reagents, waste generation, low yields, harsh reaction conditions and many more.

The designing of mild, straightforward, and environmentally friendly methods of alkene oxidation has attracted continuous interest in organic chemistry. Because of these drawbacks, green oxidants i.e. O_2, H_2O_2, TBHP, and, even air have been used as green oxidants for sustainable and eco-friendly synthesis [10–15]. Recently, metal-catalyzed oxidation of alkenes has aroused as an important tool in modern organic synthesis [16–18]. Metal organic frameworks (MOFs) have porous and heterogeneous nature and are applied for the efficient catalytic synthesis in various reactions like condensations reactions, coupling reactions, Friedel–Crafts reactions, oxidations and so on [19–22]. Since this pioneering process was developed, the field has grown considerably and a range of metal-catalyzed oxidation reactions are routinely used in synthetic chemistry laboratories all over the world. In past years, scientists utilized various metals such as Pd, Mn, Mo, Se, Fe, etc. for the effective oxidation of alkenes [23–28]. The aforementioned results showed that new protocols and mechanisms of these transformations have been developed in past decades. The present chapter systematically summarizes the metal catalyzed green oxidation of alkenes and shall provide new strategies for the functionalization and transformation of alkenes along with their advantages and disadvantages.

2. Metal catalyzed oxidation of alkene

2.1 Copper catalyzed

An easy and practical strategy has been developed for the synthesis of allylic esters by the allylic oxidation of cyclic alkenes with several carboxylic acids in the presence of tert-butyl hydroperoxide as an oxidant and copper–aluminum mixed oxide as a catalyst (**Figure 1**) [29].

A copper(I)-catalyzed oxidation of alkenes into α-oxygenated ketones was introduced using molecular oxygen and N-hydroxyphthalimide (NHPI) or N-hydroxybenzotriazole (HOBt) at room temperature (**Figure 2**). The oxidation proceeded via radical forming process and the molecular oxygen got incorporated into the product, ketone [30]. The reaction worked well for both enynes and styrenes.

An effective strategy for esterification of unactivated C-H bonds (allylic and non-benzylic) of hydrocarbons using $CuBr_2$ with different carboxylic acids was developed by Zhou et al. Selectfluor (electrophilic fluorinating agent) was used as an oxidant for the reaction. The products were obtained in 17–80% yields via cross dehydrogenative coupling (**Figure 3**) [31].

The allylic oxidation of cyclohexene was catalyzed by copper–bisoxazoline employing copper complexes $Cu(MeCN)_4PF_6$, $(CuOTf)_2{\cdot}PhH$, and $Cu(OTf)_2$ in acetonitrile or acetone as the solvents to obtain benzoic esters. Tert-butyl peroxybenzoate was used as an oxidant. The products were obtained in high yields with enantiomeric excess (<68% ee) (**Figure 4**) [32].

A study has been performed for the effect of acid and base on the oxidation of alkenes using palladium catalyst and O_2 as a sole oxidant in water as a solvent. The dihydroxylation occurred upon the oxidation in the presence of base while in the acidic medium, aldehydes and ketones were obtained by oxidative cleavage (**Figure 5**) [33].

A bis(sulfoxide) palladium catalyst (White catalyst) was applied to promote an intramolecular oxidative allylic cyclization to form biologically relevant tetracyclic flavonoids in the presence of benzoquinone as an oxidant (**Figure 6**) [34]. The reaction mechanism involved C-H activation and C-O bond formation. Both electron withdrawing and electron donating substituents provided products in high yields.

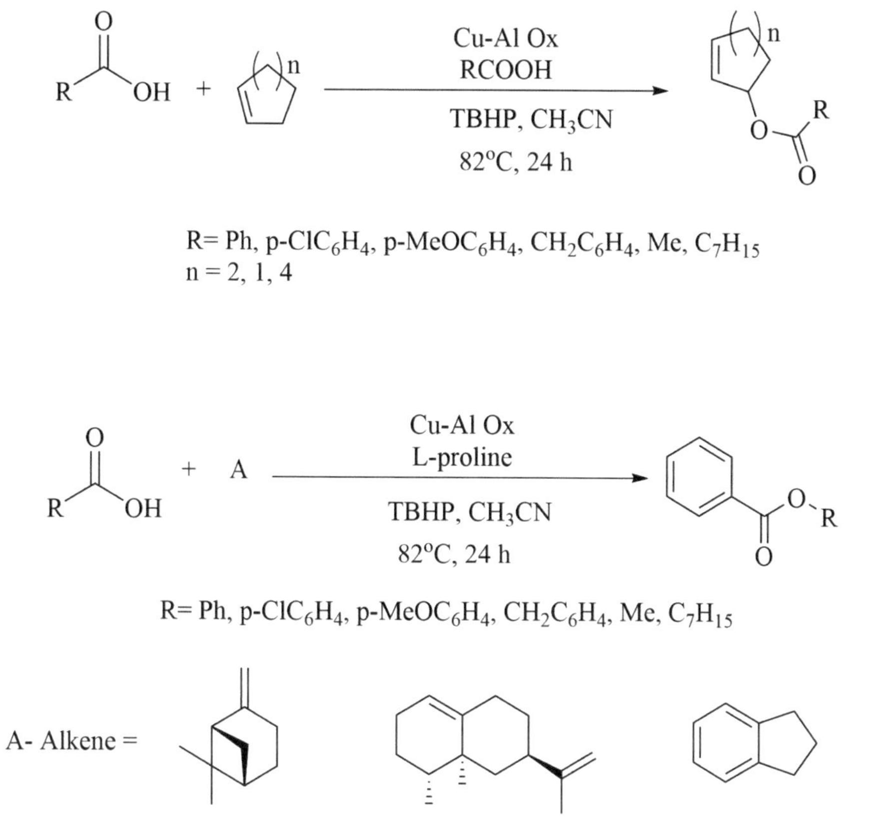

Figure 1.
Allylic oxidation of alkenes catalyzed by copper–aluminum mixed oxide

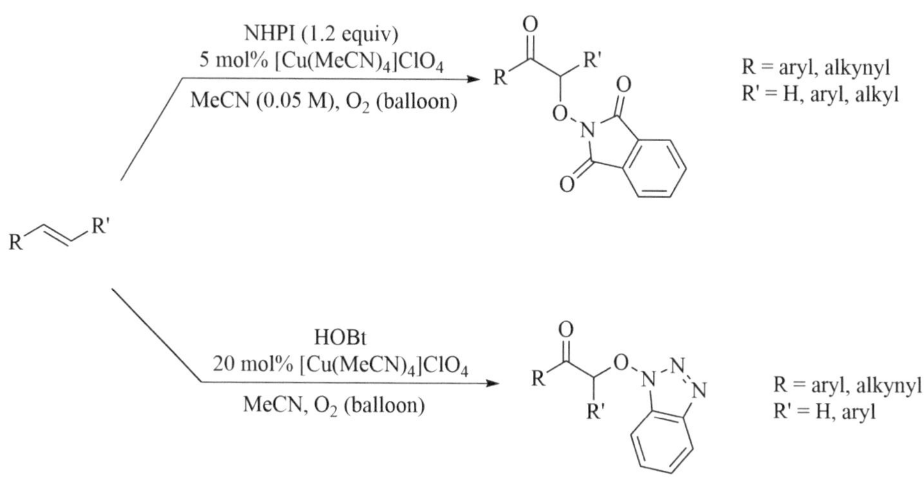

Figure 2.
Copper catalyzed oxidation of alkenes

Numerous catalytic systems have been developed for the selective formation of either linear or branched allylic oxidation products. The linear-selective allylic oxidation of alkenes was done via thioether ligand (**Figure** 7, **method a**). Furthermore, the scope of the reaction was similar to the method given firstly, by White. The reaction time was found slightly lesser in this method [35].

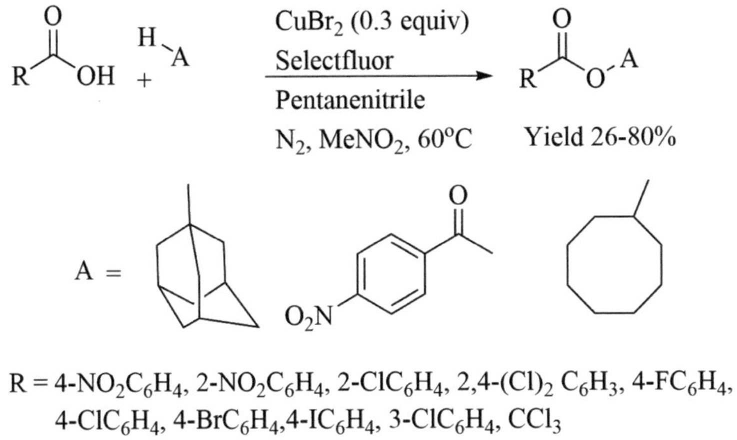

Figure 3.
Oxidation of cyclohexene using selectfluor

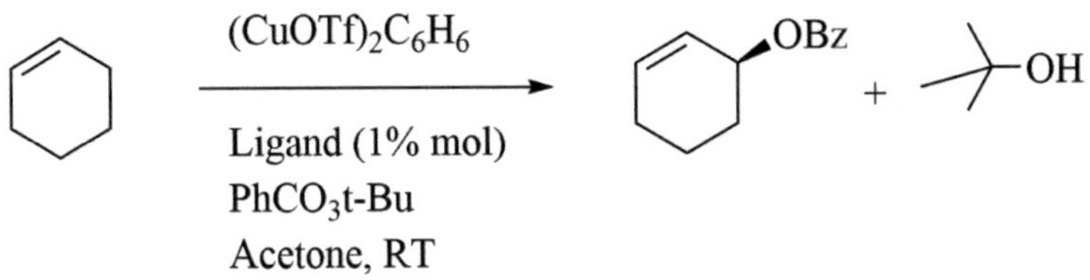

Figure 4.
Allylic oxidation of cyclohexene using $(CuOTf)_2C_6H_6$ palladium-catalyzed

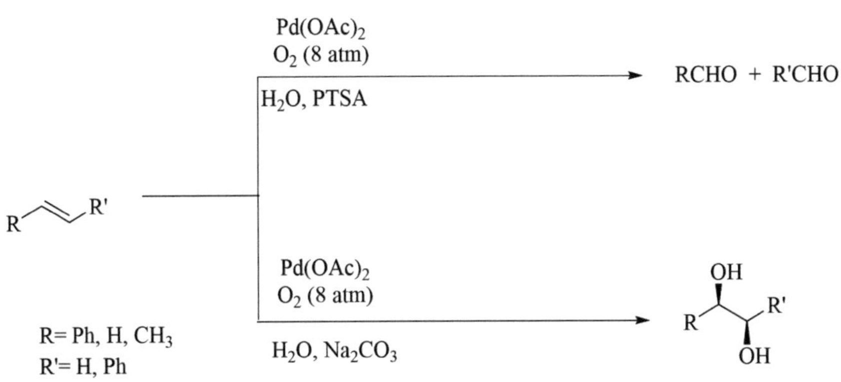

Figure 5.
Palladium-catalyzed direct oxidation of alkenes with molecular oxygen

The simple and inexpensive method was used for the synthesis of tetrahydrothiophenes which was highly active and linear-selective ligand using palladium–sulfide as catalyst (**Figure 7, method b**) [36].

The palladium–sulfide catalytic system was developed to achieve aerobic turnover with 4,5- diazafluorenone ligand (**Figure 7, method c**). This system proved to be highly selective for linear allylic acetates with good yields and functional group tolerance [37].

Water, as a nucleophile was used in the allylic oxidation of alkenes to allylic alcohols using palladium chloride as a catalyst. Water and 2,3-dichloro-5,6-dicyano-1,4-benzoquinone (DDQ) was used as a co-oxidant in the reaction (**Figure 8**) [38].

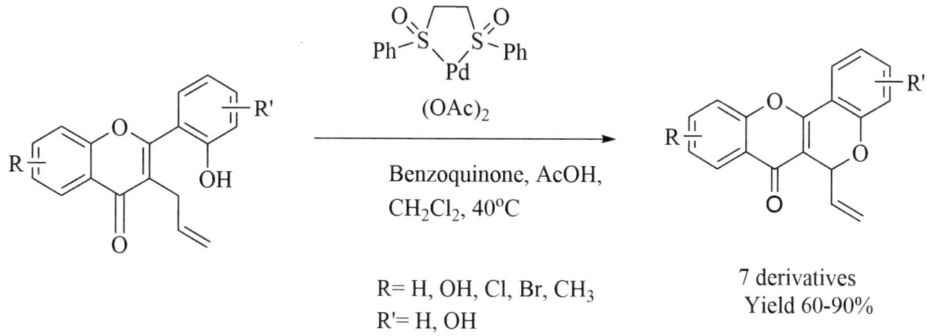

Figure 6.
Synthesis of tetracyclic flavonoids by intramolecular oxidation allylic cyclization

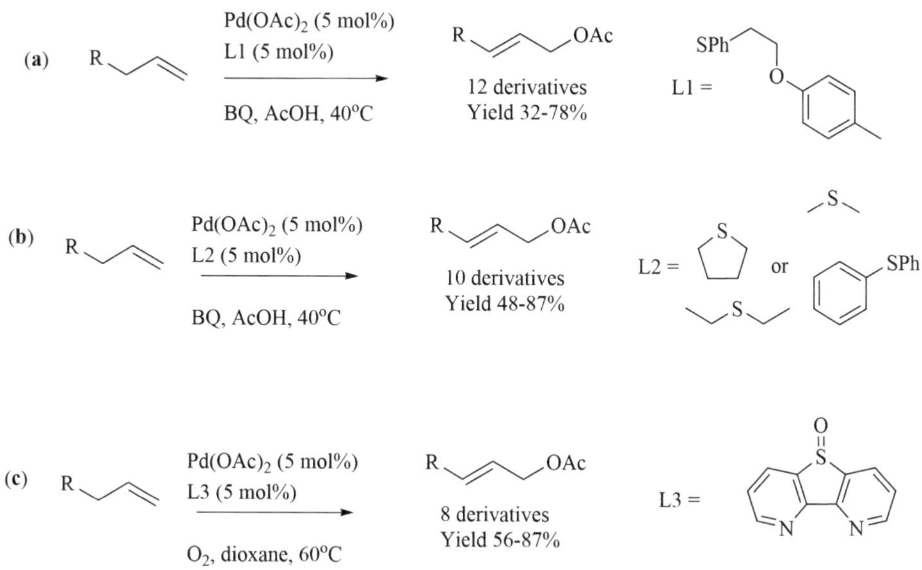

Figure 7.
Catalyst system for liner-selective allylic oxidation.

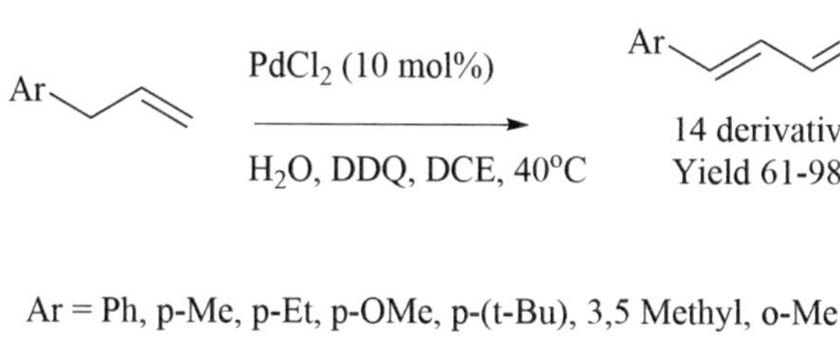

Figure 8.
Synthesis of alkenyl aldehydes by allylic oxidation with water.

The obtained products were readily oxidized to (E)-alkenyl aldehydes in excellent yields using a second eq. of DDQ.

The allylic oxidation of alkenes with polyfluorobenzenes as nucleophiles was performed. The dehydrogenative coupling was achieved in the presence of silver additives to give the corresponding linear functionalised products in good yields (**Figure 9**) [39].

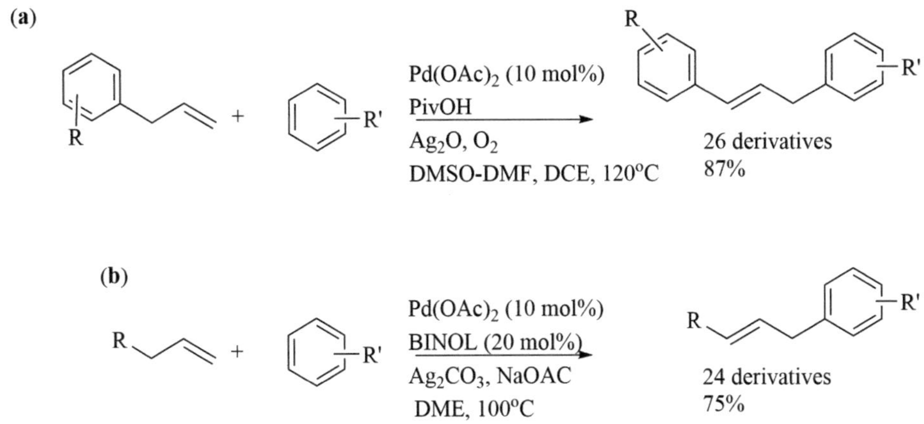

R'= 2,3,4, 5, 6-Pentafluoro; R= H, 4-Cl, 4-OMe, 2-N(Me)$_2$, 2-Me, Thiophene, Naphthyl, benzo[*d*][1,3]dioxole
R'= 2, 3, 5, 6-Tetrafluoro-4-OMe; H, 4-Ph, 4-CF$_3$,
R=H; R' = 2,3,5,6-tetrafluoro; 2, 5, 6-trilfuoro; 2,6-difluoro; 2,4,5,6-tetrafluoro-4-CF$_3$

Figure 9.
Allyic arylation with polyfluorobenzenes in the presence of silver additives.

The oxidation of unactivated alkenes resulted in a mixture of desired allylic arenes and Heck type products. A wide range of polyhalogenated nucleophiles was exemplified.

2.2 Iron catalyzed

Various iron catalysts were designed and employed for the oxidation of dienones using hydrogen peroxide as an oxidant and ethylhexanoic acid as an additive, necessary for controlling activation of the H_2O_2 at -30°C for 30 min. To optimize the reaction conditions, various solvents were examined like water, acetonitrile, THF, DMF, and toluene. Acetonitrile with 1–3 mol% of catalyst gave the best results for the reaction (**Figure 10**) [40].

A bifunctional iron nanocomposite catalyst was prepared and doped with N and P and employed for the direct oxidation of alkenes into ketones or 1,2-diketones *via* formation of an epoxide intermediate in the presence of TBHP as an oxidant in aqueous condition at mild temperature (**Figure 11**) [41].

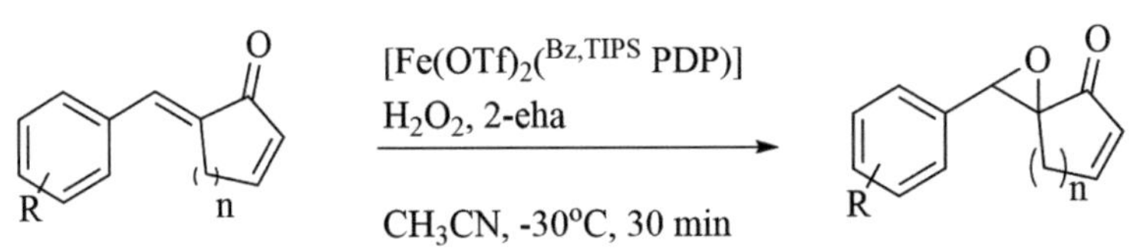

n = 1 or 2

R= H, 4-Cl, 4-F, 3-Cl, 3-F

Figure 10.
Oxidation of alkene using [Fe(OTf)$_2$(^{Bz,TIPS}PDP)].

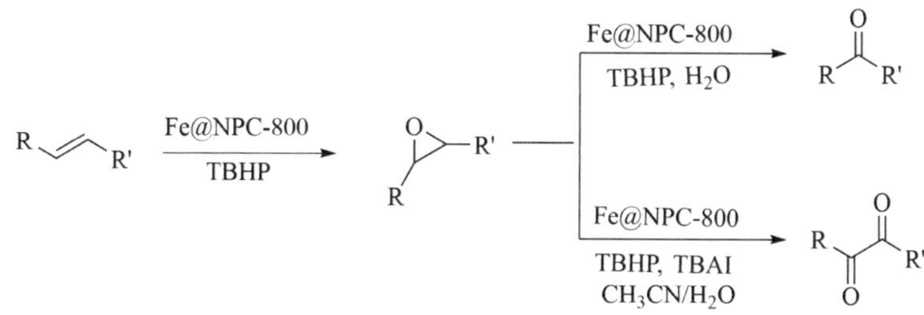

R = 4-F C_6H_4 4-Cl C_6H_4, 4-Br C_6H_4, 3-Cl C_6H_4, 4-CN C_6H_4, 3-CN C_6H_4, 4-OMe C_6H_4, 4,5-$(OMe)_2$ C_6H_3, 4-$CF_3C_6H_4$, 4-NH_2 C_6H_4
R' = C_6H_5

Figure 11.
Oxidation of alkenes by bifunctional iron nanocomposite catalyst.

2.3 Ruthenium catalyzed

A convenient and effective method was developed for the oxidation of alkenes into α-diketones using ruthenium complex as a catalyst and TBHP as an oxidant. This pathway has several advantages like no requirement of additional ligands, highly functional group tolerance, mild reaction conditions and short reaction time (**Figure 12**) [42].

2.4 Chromium-catalyzed

Mesoporous chromium terephthalate MIL- 101 was employed as a heterogeneous catalyst for the selective allylic oxidation of various alkenes in the presence of tert-butyl hydroperoxide as an oxidizing agent and provided α,β- unsaturated ketones. The catalyst MIL-101 was stable to chromium leaching and separated easily through the filtration and reused up to several cycles without any loss of catalytic activity [43].

The catalytic activities of Fe-MIL-101 and Cr-MIL-101 metal–organic frameworks were studied for the solvent-free oxidation of cyclohexene and α-pinene in the presence of molecular oxygen. Both catalysts were found effective for alkene oxidation at moderate temperature (40–60°C) and yielded allylic oxidation products. The Cr-MIL-101 afforded predominantly α,β-unsaturated ketones while Fe-MIL-101 produced mainly 2-cyclohexene-1-ol at 40°C after allylic oxidation of alkenes (**Figure 13**) [44].

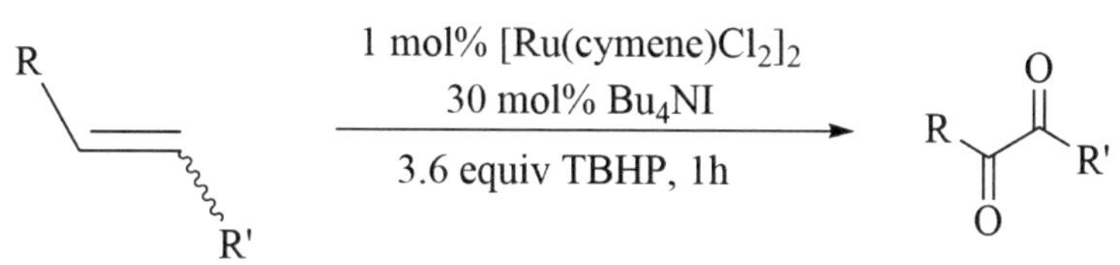

Figure 12.
Ruthenium-catalyzed oxidation of alkenes at room temperature

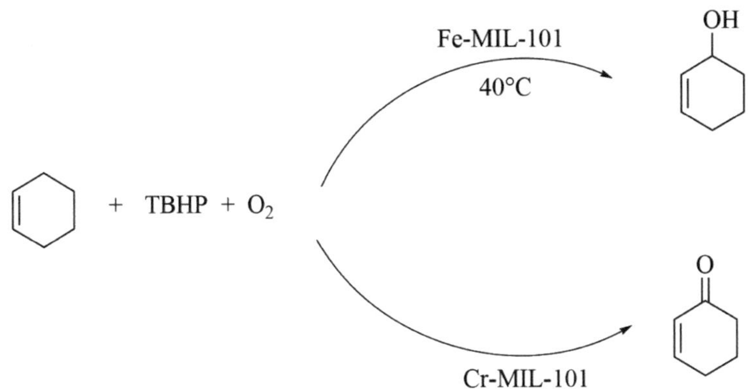

Figure 13.
Oxidation of alkenes using Fe-MIL-101 and Cr-MIL-101

2.5 Nickel catalyzed

An efficient pathway was explored for the synthesis of cyclic sulfamides through the selective oxidation of alkenes catalyzed by $NiCl_2$ or $[Ni(acac)_2]$ in the mixture of DMF and sodium acetate at 40°C in inert conditions (**Figure 14**) [45].

2.6 Selenium catalyzed

Selenium can be used as a dopant to increase the surface area and total mesoporous volume of polymeric carbon nitride through its sublimation during the process of calcination. Shi and coworkers synthesized Se-doped polymeric carbon nitride (Se/PCN) from the calcination of melamine with Se powder and used it as a catalyst for the oxidation of β ionones in the presence of visible light (**Figure 15**) [46]. In place of H_2O_2 reagent, the authors employed water as the oxidant precursor that could produce the H_2O_2 oxidant via PCN-catalyzed photolysis of water [47, 48]. Catalyzed by doped Se, β-ionone could be oxidized into epoxide in moderate yield. The use of visible light as sustainable energy to drive the reaction is the advantage of the method.

Recently, selenium based catalysts were applied for the oxidation reactions under relatively mild conditions. The authors employed H_2O_2 as the green oxidant and the transition metal-free conditions avoided the metal residues in the product. Different from the reported selenium catalyzed oxidation reactions using diaryl diselenide or aryl selenic acid catalyst, 5 mol% of dialkyl diselenide, such as $(PhCH_2Se)_2$, $(n\text{-}C_4H_9Se)_2$, or $(c\text{-}C_6H_{11}Se)_2$ was used as the pre-catalyst, converting the alkene substrates into the related carbonyls in 30–74% yields in EtOH at 80–120°C (**Figure 16**) [49].

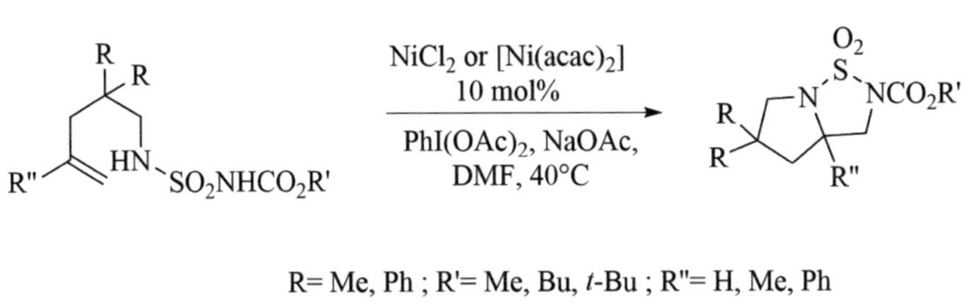

Figure 14.
Nickel catalyzed oxidation of alkenes

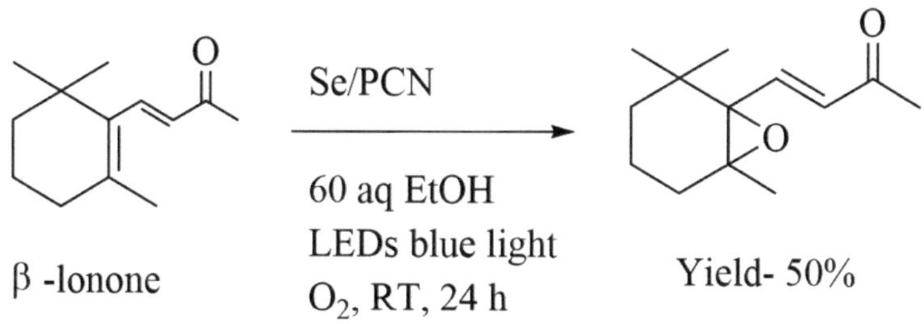

Figure 15.
Oxidation of β-Ionone in the presence of Se/PCN

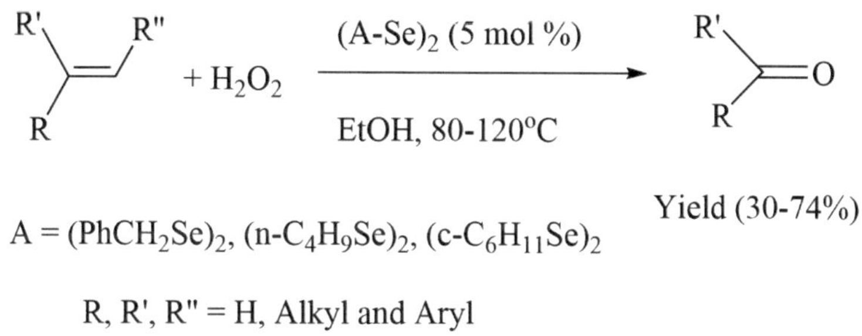

Figure 16.
Oxidation of alkene in the presence $(RSe)_2$ with H_2O_2

Selenization of dihalohydrocarbon with NaHSe (prepared via reduction of Se powder with $NaBH_4$ in situ) afforded polyselenide in just one step and it was found to be an efficient catalyst for alkene oxidative cracking reactions under mild conditions (**Figure 17**) [50].

The selenization of Fe_2O_3 afforded another convenient access to heterogeneous Se catalyst [51]. The developed method may be used at the industrial scale due to cheap and abundant nature of the support Fe_2O_3. The reaction mechanism involved attack of HSe^- anion on the iron positive centre of Fe = O and uploading of Se to produce Se/Fe_2O_3. It efficiently catalyzed the oxidative of alkenes in dimethyl

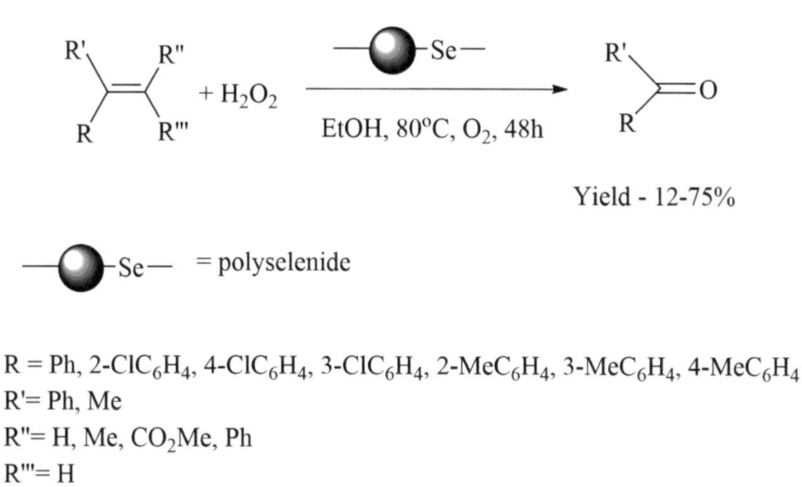

Figure 17.
Oxidation of alkenes using polyselenide as a highly efficient catalyst

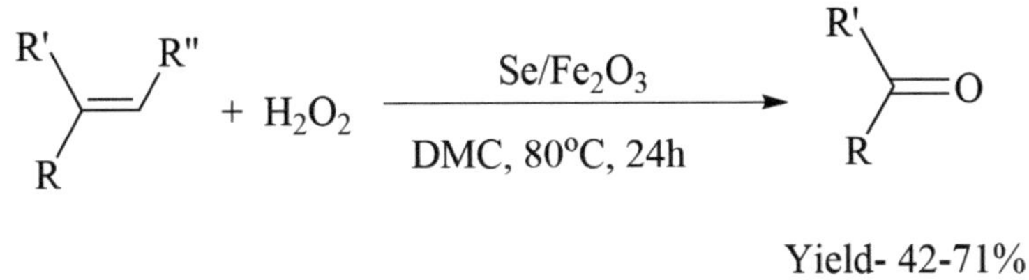

Figure 18.
Oxidation of alkenes in the presence of Se/Fe2O3

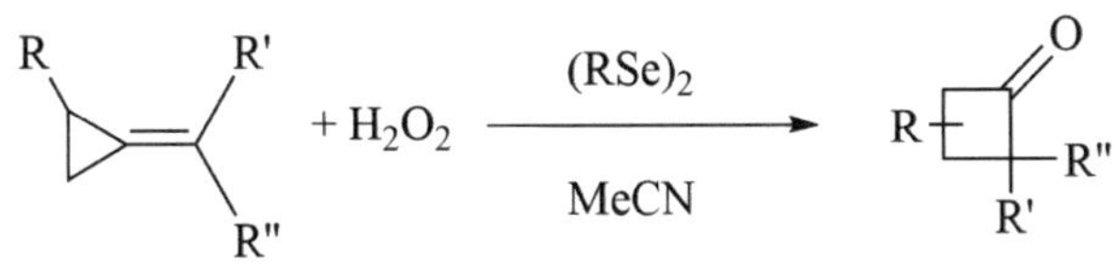

Figure 19.
Se-catalyzed oxidation of methylenecyclopropanes

carbonate (DMC) under mild reaction conditions (**Figure 18**). The synergistic effect of Se with Fe in the material, Se/Fe_2O_3 enhanced its catalytic activity [52].

The Se-catalyzed oxidation of methylenecyclopropanes (MCPs) generated oxidative ring expansion products, cyclobutanones instead of epoxides (**Figure 19**). This reaction may afford a direct access to cyclobutanone intermediates. The driving force for the conversion is the high ring strain of MCPs [53].

3. Metal based nano-particle catalyzed oxidation of alkenes

An eco-benign pathway was developed for the oxidation of alkenes to ketones using water as an oxidant and catalyst combination of dearomatized acridine-based PNP-Ru complex and indium(III) triflate. The present protocol showed good functional group tolerance. The pathway was highly advantageous over Wacker-type oxidation as no chemical oxidant was used and only hydrogen gas was liberated as the byproduct (**Figure 20**) [54].

A variety of mesoporous nanocomposites comprising of zirconium oxide and Keggin 12-phosphomolybdic acid compounds (ZrO_2–PMA) was synthesized through the application of evaporation-induced cooperative assembly method and were employed as an efficient catalyst for the oxidation of alkenes in the presence of hydrogen peroxide as an oxidant. ZrO_2 impregnated with nearly 37 wt% PMA exhibited significant activity for the oxidation of alkenes (**Figure 21**) [55].

Cobalt ferrite magnetic nanoparticles ($CoFe_2O_4$ NPs) were used as effective catalysts for the selective oxidation of alkenes into related aldehydes or epoxides in the presence of t-BuOOH as an oxidant and 1, 2-dichloroethane as a solvent. The catalyst was easily separated from the reaction medium using a magnet, and reused up to five consecutive cycles without any loss of activity (**Figure 22**) [56].

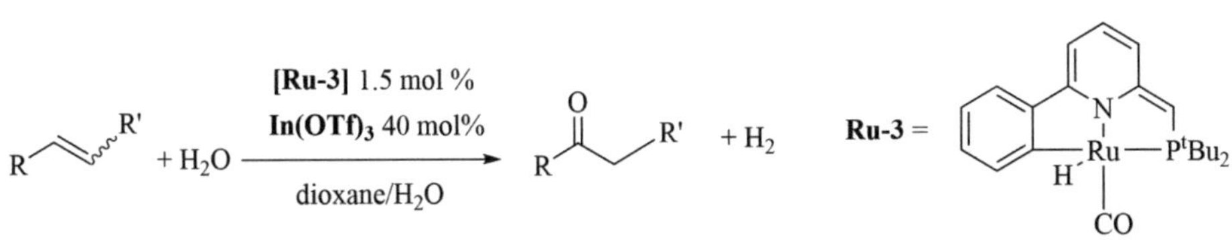

Figure 20.
Oxidation of alkenes by water

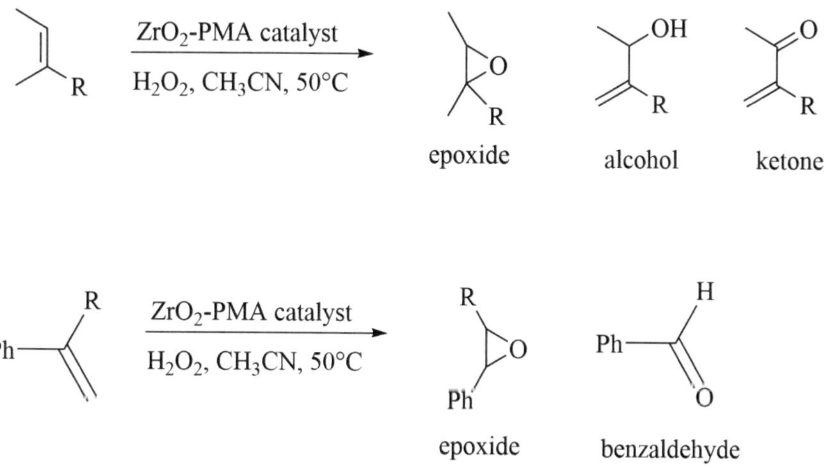

Figure 21.
Oxidation of alkenes by mesoporous zirconia-polyoxometalate nanocomposite materials

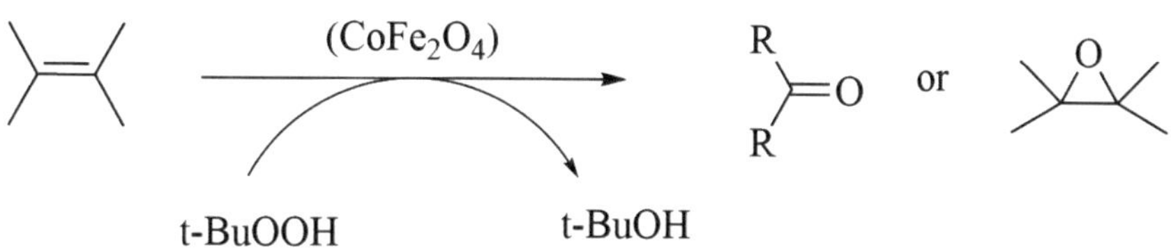

Figure 22.
Magnetic cobalt ferrite NPs as an efficient catalyst for oxidation of alkenes

A green method was developed for the oxidation of alkenes using chiral dicationic bisguanidinium in the presence of potassium permanganate at 60°C for 12–36 h. (**Figure 23**) [57].

An eco-friendly and highly efficient protocol was developed for the epoxidation of alkenes with m-chloroperbenzoic acid in the presence of Fe_3O_4/SiO_2 as a heterogeneous catalyst in DCM as solvent at mild reaction conditions in 4 h at RT. In addition, the prepared heterogeneous catalyst has some advantages such as easy preparation, reusability, handling and recovery, long-term stability as well as excellent yields (90–97%) in shorter reaction time (**Figure 24**) [58].

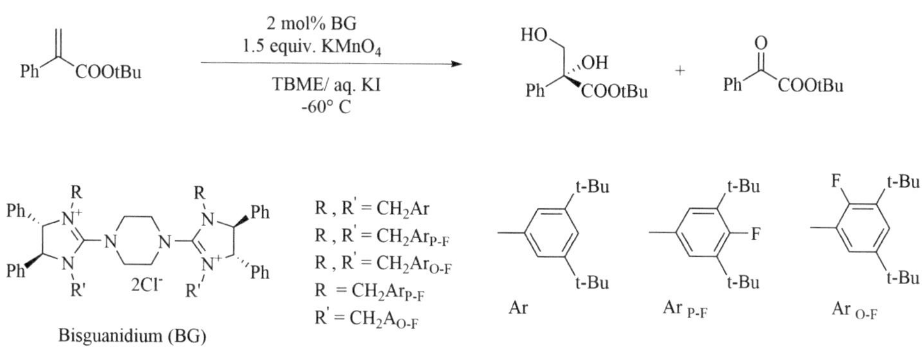

Figure 23.
Enantioselective oxidation of alkenes with potassium permanganate catalyzed by chiral Dicationic Bisguanidinium

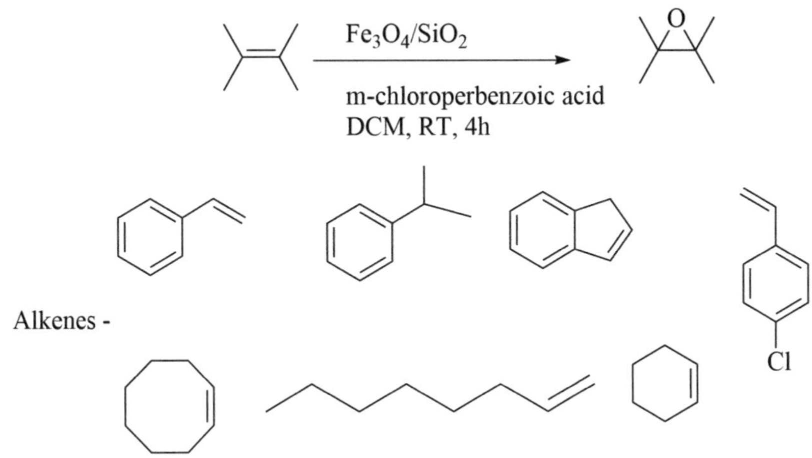

Figure 24.
Oxidation of alkenes in the presence of Fe_3O_4/SiO_2

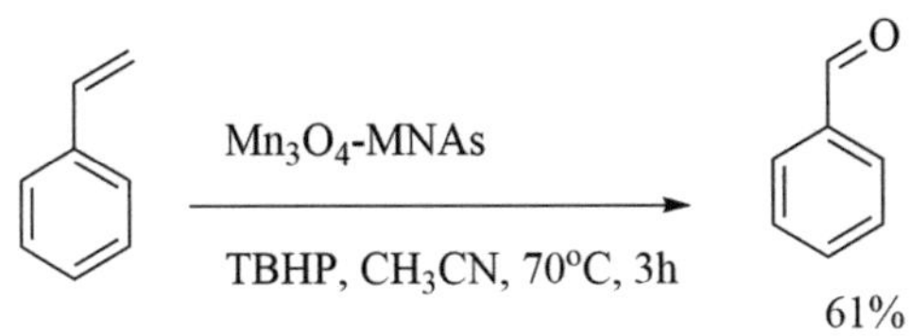

Figure 25.
Oxidation of styrene in the presence of Mn_3O_4-MNAs with TBHP

Oxidation of styrene was done in the presence of mesoporous Mn_3O_4 nanoparticle (0.04 mmol) as catalyst and tert butyl hydroperoxide as mild oxidant with CH_3CN at 70°C for 3 h. Novel, simple, and high yield (67–91%) of products are the advantages of Mn_3O_4-MNAs. The catalyst was recycled and reused up to five runs without a significant diminution in the catalytic efficiency (**Figure 25**) [59].

4. Applications

In this section, several cases were found that employed the copper-catalyzed allylic oxidation of olefins as an important step in the syntheses of synthetic products and pharmaceuticals. Some of the examples were exemplified by Nakamura and Nakada in their review on allylic oxidation reactions for the synthesis of natural products [60].

In 2009, Hayashi et al. produced the enantioselective synthesis of an O-protected 2-deoxystreptamine precursor, an amino-glycoside antibiotic, the same copper-catalyzed asymmetric desymmetrization procedure for the preparation of a chiral intermediate in the total synthesis of (−)-oseltamivir phosphate [61, 62].

The synthesis of steroids is one field where copper-catalyzed allylic oxidation has been more successful. The relatively inert steroid framework usually yielded the corresponding enone instead of an allylic alcohol. These types of phytosterol could have vivid applications in the field of medicine owing to their ability of modulating mitochondrial activity and their anti-inflammatory properties. Several examples of Kharasch–Sosnovsky reactions for the synthesis of steroids are available in the literature [63].

Barbacenic acid, a novel bisnorditerpene isolated from *Barbacenia flava*, was synthesized by a highly stereocontrolled route. The authors reported the allylic oxidation on the silyl ether intermediate using TBHP in the presence of CuBr in benzene at 45°C under mild reaction conditions in 60% yield [64]. Moreover, oxidized alkenes were used in the synthesis of fine chemicals [65], arene reduction [66], and water repellency [67].

5. Conclusion

This chapter summarizes the recent strategies of the oxidation of alkenes *via* metal-based catalytic system. Various innovations have been implemented from time to time to increase the efficiency and sustainability of the protocols. This literature study demonstrated the potential of metal-based catalysts and nano-particle metal based catalysts in the oxidation of alkenes to produce different chemical entities such as carbonyl, 1,2-diol and allylic compounds. However, some challenges like low commercial applicability, less efficiency, low chemoselectivity control, over-oxidation, lack of knowledge and deeper understanding of mechanistic pathways are still present in this field. In this context, the improvement in industrially viability, efficiency and sustainability of synthetic pathways require innovation in both capability and strategy. The knowledge gained for the oxidation of alkenes from these studies along with their pros and cons will continue to enable advances in this area, which in turn shall lead to the advancement of new approaches.

Acknowledgements

The authors are grateful to Department of Chemistry, MLSU, Udaipur (Raj.), India for providing necessary library facilities. A. Sethiya, N. Sahiba and P. Teli are thankful to UGC-MANF and CSIR for providing the financial support.

Conflict of interest

The authors declare no conflict of interest.

Author details

Jay Soni, Pankaj Teli, Nusrat Sahiba, Ayushi Sethiya and Shikha Agarwal*
Synthetic Organic Chemistry Laboratory, Department of Chemistry, MLSU, Udaipur, Rajasthan, India

*Address all correspondence to: shikhaagarwal@mlsu.ac.in

References

[1] Yang KS, Gurak Jr JA, Liu Z, Engle KM. Catalytic, regioselective hydrocarbofunctionalization of unactivated alkenes with diverse C–H nucleophiles. J. Am. Chem. Soc. 2016;138(44):14705-14712.

[2] Kolb HC, VanNieuwenhze MS, Sharpless KB. Catalytic asymmetric dihydroxylation. Chem. Rev. 1994 Dec;94(8):2483-2547.

[3] Zaitsev AB, Adolfsson H. Recent developments in asymmetric dihydroxylations. Synthesis. 2006;2006(11):1725-1756.

[4] Bataille CJ, Donohoe TJ. Osmium-free direct syn-dihydroxylation of alkenes. Chem. Soc. Rev. 2011;40(1):114-128.

[5] Ogino T, Mochizuki K. Homogeneous permanganate oxidation in non-aqueous organic solution. Selective oxidations of olefins into 1, 2-diols or aldehydes. Chem. Lett. 1979;8(5):443-446.

[6] Wiberg KB, Saegebarth KA. The mechanisms of permanganate oxidation. IV. Hydroxylation of olefins and related reactions. J. Am. Chem. Soc. 1957;79(11):2822-2824.

[7] Freeman F, Yamachika NJ. Chromyl chloride oxidations. IV. Kinetics and mechanism of the addition to styrene. J. Am. Chem. Soc. 1970;92(12):3730-3733.

[8] Berkowitz LM, Rylander PN. Use of ruthenium tetroxide as a multi-purpose oxidant. J. Am. Chem. Soc. 1958;80(24):6682-6684.

[9] Pappo R, Allen, Jr D, Lemieux R, Johnson WS. Osmium tetroxide-catalyzed periodate oxidation of olefinic bonds. J. Org. Chem. 1956;21(4):478-479.

[10] Gao G, Han J, Yu L, Xu Q. Organoselenium-catalyzed polymerization of aniline with hydrogen peroxide as oxidant. Synlett. 2019;30(14):1703-1707.

[11] Fang W, Lin X, Cheng S, Qing X, Jiejun H, Lei Y. Investigation on Preparation of p-Benzoquinone through the Organoselenium-Catalyzed Selective Oxidation of Phenol. Chin. J. Org. Chem. 2017;37(8):2115-2118.

[12] Depken C, Krätzschmar F, Rieger R, Rode K, Breder A. Photocatalytic aerobic phosphatation of alkenes. Angew. Chem. Int. Ed. 2018;57(9):2459-2463.

[13] Rode K, Palomba M, Ortgies S, Rieger R, Breder A. Aerobic Allylation of Alcohols with Non-Activated Alkenes Enabled by Light-Driven Selenium-π-Acid Catalysis. Synthesis. 2018;50(19):3875-3885.

[14] Jing X, Wang T, Ding Y, Yu L. A scalable production of anisonitrile through organoselenium-catalyzed dehydration of anisaldoxime under solventless conditions. App. Catal. A: Gen. 2017;541:107-111.

[15] Zhang X, Sun J, Ding Y, Yu L. Dehydration of aldoximes using PhSe (O) OH as the pre-catalyst in air. Org. lett. 2015;17(23):5840-5842.

[16] Ballistreri FP, Gangemi C, Pappalardo A, Tomaselli GA, Toscano RM, Trusso Sfrazzetto G. (Salen) Mn (III) catalyzed asymmetric epoxidation reactions by hydrogen peroxide in water: a green protocol. Int. J. Mol. Sci. 2016;17(7):1112.

[17] Kashani SH, Moghadam M, Tangestaninejad S, Mirkhani V, Mohammadpoor-Baltork I. Ruthenium nanoparticles immobilized on nano-silica functionalized with thiol-based dendrimer: a nanocomposite material for oxidation of alcohols and

epoxidation of alkenes. Catal. Lett. 2018;148(4):1110-1123.

[18] Zhu Y, Wang Q, Cornwall RG, Shi Y. Organocatalytic asymmetric epoxidation and aziridination of olefins and their synthetic applications. Chem. Rev. 2014;114(16):8199-8256.

[19] Anbu N, Dhakshinamoorthy A. Cu3 (BTC) 2 catalyzed oxidation of silane to silanol using TBHP or water as oxidants. App. Catal. A: Gen. 2017;544:145-153.

[20] Kholdeeva OA, Skobelev IY, Ivanchikova ID, Kovalenko KA, Fedin VP, Sorokin AB. Hydrocarbon oxidation over Fe-and Cr-containing metal-organic frameworks MIL-100 and MIL-101–a comparative study. Catal. Today. 2014;238:54-61.

[21] Skobelev IY, Sorokin AB, Kovalenko KA, Fedin VP, Kholdeeva OA. Solvent-free allylic oxidation of alkenes with O2 mediated by Fe-and Cr-MIL-101. J. Catal. 2013;298:61-69.

[22] Torbina VV, Ivanchikova ID, Kholdeeva OA, Skobelev IY, Vodyankina OV. Propylene glycol oxidation with tert-butyl hydroperoxide over Cr-containing metal-organic frameworks MIL-101 and MIL-100. Catal. Today. 2016;278:97-103.

[23] Moiseev II, Vargaftik MN. Allylic oxidation of alkenes with palladium catalysts. Coord. Chem. Rev. 2004;248(21-24):2381-2391.

[24] Nodzewska A, Wadolowska A, Watkinson M. Recent advances in the catalytic oxidation of alkene and alkane substrates using immobilized manganese complexes with nitrogen containing ligands. Coord. Chem. Rev. 2019;382:181-216.

[25] Shen Y, Jiang P, Wai PT, Gu Q, Zhang W. Recent progress in application of molybdenum-based catalysts for epoxidation of alkenes. Catalysts. 2019;9(1):31.

[26] Cao H, Qian R, Yu L. Selenium-catalyzed oxidation of alkenes: insight into the mechanisms and developing trend. Catal. Sci. Technol. 2020;10(10):3113-3121.

[27] Olivo G, Cussó O, Borrell M, Costas M. Oxidation of alkane and alkene moieties with biologically inspired nonheme iron catalysts and hydrogen peroxide: from free radicals to stereoselective transformations. J. Biol. Inorg. Chem. 2017;22(2):425-452.

[28] Borrell M, Costas M. Greening oxidation catalysis: iron catalyzed alkene syn-dihydroxylation with aqueous hydrogen peroxide in green solvents. ACS Sustain. Chem. Eng. 2018;6(7):8410-8416.

[29] García-Cabeza AL, Marín-Barrios R, Moreno-Dorado FJ, Ortega MJ, Massanet GM, Guerra FM. Allylic oxidation of alkenes catalyzed by a copper–aluminum mixed oxide. Org. Lett. 2014;16(6):1598-1601.

[30] Andia AA, Miner MR, Woerpel KA. Copper (I)-catalyzed oxidation of alkenes using molecular oxygen and hydroxylamines: synthesis and reactivity of α-oxygenated ketones. Org. Lett. 2015;17(11):2704-2707.

[31] Zhou J, Jin C, Li X, Su W. Copper-catalyzed oxidative esterification of unactivated C (sp 3)–H bonds with carboxylic acids via cross dehydrogenative coupling. RSC Adv. 2015;5(10):7232-7236.

[32] Alvarez LX, Christ ML, Sorokin AB. Selective oxidation of alkenes and alkynes catalyzed by copper complexes. Appl. Catal. A: Gen. 2007;325(2):303-308.

[33] Wang A, Jiang H. Palladium-catalyzed direct oxidation of alkenes

with molecular oxygen: general and practical methods for the preparation of 1, 2-diols, aldehydes, and ketones. J. Org. Chem. 2010;75(7):2321-2326.

[34] Ayyagari N, Belani JD. letter Synthesis of Tetracyclic Flavonoids via Palladium-Catalyzed Intramolecular Oxidative Cyclization. Synlett. 2014;25:2350-4.

[35] Henderson WH, Check CT, Proust N, Stambuli JP. Allylic oxidations of terminal olefins using a palladium thioether catalyst. Org. Lett. 2010;12(4):824-827.

[36] Le CC, Kunchithapatham K, Henderson WH, Check CT, Stambuli JP. A Survey of Sulfide Ligands for Allylic C-H Oxidations of Terminal Olefins. Chem.–A Eur. J. 2013;19(34): 11153-11157.

[37] Campbell AN, White PB, Guzei IA, Stahl SS. Allylic C– H acetoxylation with a 4, 5-diazafluorenone-ligated palladium catalyst: A ligand-based strategy to achieve aerobic catalytic turnover. J. Am. Chem. Soc. 2010;132(43):15116-15119.

[38] Chen H, Jiang H, Cai C, Dong J, Fu W. Facile synthesis of (E)-alkenyl aldehydes from allyl arenes or alkenes via Pd (II)-catalyzed direct oxygenation of allylic C– H bond. Org. Lett. 2011;13(5):992-994.

[39] Jiang H, Yang W, Chen H, Li J, Wu W. Palladium-catalyzed aerobic oxidative allylic C–H arylation of alkenes with polyfluorobenzenes. Chem. Commun. 2014;50(54): 7202-7204.

[40] Cusso O, Cianfanelli M, Ribas X, Klein Gebbink RJ, Costas M. Iron catalyzed highly enantioselective epoxidation of cyclic aliphatic enones with aqueous H_2O_2. J. Am. Chem. Soc. 2016;138(8):2732-2738.

[41] Song T, Ma Z, Ren P, Yuan Y, Xiao J, Yang Y. A Bifunctional Iron Nanocomposite Catalyst for Efficient Oxidation of Alkenes to Ketones and 1, 2-Diketones. ACS Catal. 2020;10(8):4617-4629.

[42] Chen S, Liu Z, Shi E, Chen L, Wei W, Li H, Cheng Y, Wan X. Ruthenium-catalyzed oxidation of alkenes at room temperature: a practical and concise approach to α-diketones. Org. Lett. 2011;13(9):2274-2277.

[43] Maksimchuk NV, Kovalenko KA, Fedin VP, Kholdeeva OA. Heterogeneous Selective Oxidation of Alkenes to α, β-Unsaturated Ketones over Coordination Polymer MIL-101. Adv. Synth. Catal. 2010;352(17):2943-2948.

[44] Skobelev IY, Sorokin AB, Kovalenko KA, Fedin VP, Kholdeeva OA. Solvent-free allylic oxidation of alkenes with O_2 mediated by Fe-and Cr-MIL-101. J. Catal. 2013;298:61-69.

[45] Muniz K, Streuff J, Hövelmann CH, Nunez A. Exploring the Nickel-Catalyzed Oxidation of Alkenes: A Diamination by Sulfamide Transfer. Angew. Chem. Int. Ed. 2007;46(37):7125-7127.

[46] Li H, Cao H, Chen T, Zhang X, Shi Y. Selenium-incorporated polymeric carbon nitride for visible-light photocatalytic regio-specific epoxidation of β-ionone. Mol. Catal. 2020;483:110715.

[47] Ling X, Xu Y, Wu S, Liu M, Yang P, Qiu C, Zhang G, Zhou H, Su C. A visible-light-photocatalytic water-splitting strategy for sustainable hydrogenation/deuteration of aryl chlorides. Sci. China Chem. 2020;63(3):386-392.

[48] Fan X, Yao Y, Xu Y, Yu L, Qiu C. Visible-Light-Driven Photocatalytic Hydrogenation of Olefins Using Water

as the H Source. ChemCatChem. 2019 Jun 6;11(11):2596-2599.

[49] Wang T, Jing X, Chen C, Yu L. Organoselenium-Catalyzed Oxidative C-C Bond Cleavage: A Relatively Green Oxidation of Alkenes into Carbonyl Compounds with Hydrogen Peroxide. J. Org, Chem. 2017;82(18):9342-9349.

[50] Yu L, Cao H, Zhang X, Chen Y, Yu L. Concise synthesis of polyselenides: efficient catalysts for the oxidative cracking reaction of alkenes allowing the utilization of O_2 as a partial oxidant under mild conditions. Sustain. Energ. Fuels. 2020;4:730-736.

[51] Liu C, Mao J, Zhang X, Yu L. Selenium-doped Fe_2O_3-catalyzed oxidative scission of C=C bond. Catal. Commun. 2020;133:105828-105834.

[52] Chen C, Zhang X, Cao H, Wang F, Yu L, Xu Q. Iron-Enabled Utilization of Air as the Terminal Oxidant Leading to Aerobic Oxidative Deoximation by Organoselenium Catalysis. Adv. Synth. Catal. 2019:361;603-608.

[53] L. Yu, M. Shi. The Construction of Molecular Complexity from Functionalized Alkylidenecyclopropanes (FACPs). Chem. Eur. J. 2019:25;7591.

[54] Tang S, Ben-David Y, Milstein D. Oxidation of Alkenes by Water with H2 Liberation. J. Am. Chem. Soc. 2020;142(13):5980-5984.

[55] Armatas GS, Bilis G, Louloudi M. Highly ordered mesoporous zirconia-polyoxometalate nanocomposite materials for catalytic oxidation of alkenes. J. Mater. Chem. 2011;21(9):2997-3005.

[56] Kooti M, Afshari M. Magnetic cobalt ferrite nanoparticles as an efficient catalyst for oxidation of alkenes. Scientia Iranica. 2012;19(6):1991-1995.

[57] Wang C, Zong L, Tan CH. Enantioselective oxidation of alkenes with potassium permanganate catalyzed by chiral dicationic bisguanidinium. J. Am. Chem. Soc. 2015;137(33):10677-10682.

[58] Nasseri MA, Allahresani A, Raissi H. Mild oxidation of alkenes catalyzed by Fe_3O_4/SiO_2 nanoparticles. React. Kinet. Mech. Catal. 2014;112(2):397-408.

[59] Skliri E, Papadogiorgakis S, Lykakis IN, Armatas GS. Mesoporous assembled Mn_3O_4 nanoparticle networks as efficient catalysts for selective oxidation of alkenes and aryl alkanes. ChemPlusChem. 2017;82(1):136-143.

[60] Nakamura A, Nakada M. Allylic oxidations in natural product synthesis. Synthesis. 2013;45(11):1421-1451.

[61] Tan Q, Hayashi M. Asymmetric desymmetrization of 4, 5-epoxycyclohex-1-ene by enantioselective allylic oxidation. Org. Lett. 2009;11(15):3314-3317.

[62] Yamatsugu K, Kamijo S, Suto Y, Kanai M, Shibasaki M. A concise synthesis of Tamiflu: third generation route via the Diels–Alder reaction and the Curtius rearrangement. Tetrahedron Lett. 2007;48(8):1403-1406.

[63] Rahman FU, Rahman AU, Tan TW. Direct Cu-Catalyzed Allylic Acetoxylation of Δ5-Steroids at 7-Position. J. Chinese Chem. Soci. 2010;57(6):1237-1242.

[64] Patin A, Kanazawa A, Philouze C, Greene AE, Muri E, Barreiro E, Costa PC. Highly Stereocontrolled Synthesis of Natural Barbacenic Acid, Novel Bisnorditerpene from Barbacenia f lava. J. Org. Chem. 2003;68(10):3831-3837.

[65] Bicchielli, D., Borguet, Y., Delaude, L., Demonceau, A., Dragutan, I.,

Dragutan, V., Sauvage, X. Recent Applications of Alkene Metathesis in Fine Chemical Synthesis. NATO Science for Peace and Security Series A: Chemistry and Biology 2010;207-274. doi:10.1007/978-90-481-3433-5_15

[66] Sawama Y, Yasukawa N, Ban K, Goto R, Niikawa M, Monguchi Y, Itoh M, Sajiki H. Stainless steel-mediated hydrogen generation from alkanes and diethyl ether and its application for arene reduction. Org. Lett. 2018;20(10):2892-2896.

[67] Liu X, Zou X, Ge Z, Zhang W, Luo Y. Novel waterborne polyurethanes containing long-chain alkanes: their synthesis and application to water repellency. RSC Adv. 2019;9(54): 31357-31369.

Chapter 3

Construction of C-N Bond *via* Visible-Light-Mediated Difunctionalization of Alkenes

Bhisma K. Patel, Tipu Alam and Amitava Rakshit

Abstract

In the last few years, the photo-redox process via single-electron transfer (SET) has received substantial attention for the synthesis of targeted organic compounds due to its environmental friendliness and sustainability. Of late visible-light-mediated difunctionalization of alkenes has gained much attention because of its step economy, which allows the consecutive installation of two functional groups across the C=C bond in a single operation. The construction of *N*-containing compounds has always been important in organic synthesis. Molecules containing C-N bonds are found in many building blocks and are important precursors to other functional groups. Meanwhile, C-N bond formation via the addition of the C=C double bond is gaining prominence. Therefore, considering the influence and synthetic potential of the C-N bond, here we provide a summary of the state of the art on visible-light-driven difunctionalizations of alkene. We hope that the construction of the C-N bond *via* visible-light-mediated difunctionalization of alkenes will be useful for medicinal and synthetic organic chemists and will inspire further reaction development in this interesting area.

Keywords: Photocatalysis, Difunctionalization, C-N Bond, Alkenes

1. Introduction

Of late, photo-redox catalysis has been utilized as a flexible and demanding synthetic protocol in the realm of modern organic chemistry due to its environmental friendliness and sustainability [1]. This visible-light-driven protocol essentially affords a large number of nitrogen centred radicals via a single electron transfer (SET) process or energy transfer process under mild reaction conditions, compared to the traditional radical reactions that use high-energy ultraviolet (UV) light or highly toxic and expensive radical initiators [2]. Therefore, visible-light-mediated photo-redox catalysis has been widely applied for the synthesis of natural products, synthetic methodologies, enantioselective catalysis, and polymerization reactions. The success of any photochemical reactions relies on the ability of photocatalysts, usually transition-metal based complexes, organic dyes or heterogeneous semiconductors which promote single-electron transfer (SET) with organic molecules upon excitation with visible light [3]. An alkene difunctionalization can introduce two functional groups in a signal operation across the double bond (**Figure 1**) [4]. In this context, the radical-mediated C-N bond formation has

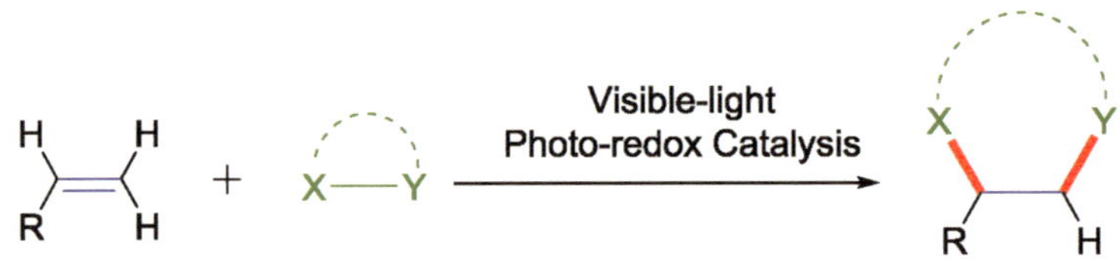

Figure 1.
Difunctionalization of alkene via visible-light photo-redox catalysis.

emerged as a powerful strategy to construct valuable molecules that have found application in different fields [5]. Therefore, this chapter focuses mainly on the difunctionalization of C=C bonds emphasizing the C-N bond formation using visible-light photo-redox catalysis.

2. Photocatalysts

Photocatalysts are organic or inorganic substances that absorb light and get excited to a higher energy level and transfer this energy to a reacting partner thereby triggering a chemical reaction. Few commonly used photocatalysts for the difunctionalization of alkenes [6] are shown in (**Figures 2** and **3**). These photocatalysts can be divided into two categories: (A) Transition-metal complexes and (B) Organic dyes.

A. **Transition-Metal Complex Photocatalysts**: The widely used visible-light-mediated photocatalyst are well-defined ruthenium (II) polypyridine complexes or Ir-cyclometalated derivatives. They facilitate redox reactions due to their ability to cause single electron transfer (SET) under a mild reaction condition in the presence of a visible light source [7].

B. **Organic Dye**: Since the beginning of organic synthesis, the formation of carbon-heteroatom bonds in a controlled and efficient manner is the heart of organic synthesis. Organic dyes using visible light has been playing a key role

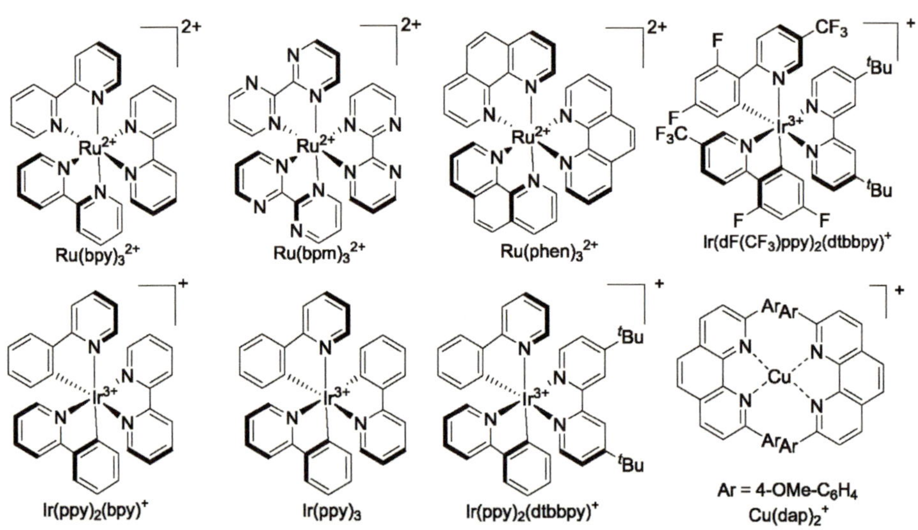

Figure 2.
Commonly used metal-based Photocatalyst for Difunctionalization.

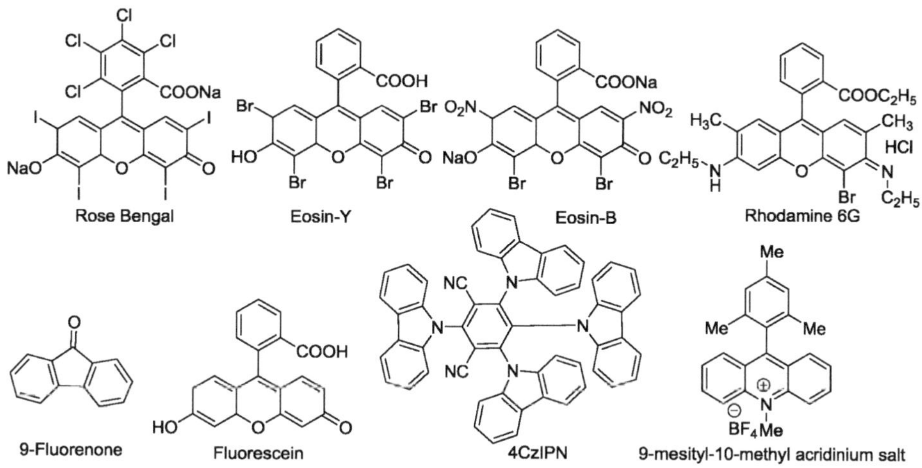

Figure 3.
Commonly used organic dye Photocatalyst for Difunctionalization of alkene.

in the formations of carbon-heteroatom bond in the difunctionalization of alkene. Compared to metal-containing photocatalyst, organic dyes require mild reaction conditions as it uses visible light of low power source and therefore the number of organic dyes have been employed as photo redox catalysts [8].

3. General mechanism

In general, at first, a photocatalyst (P.C) is converted to its excited state (*P.C) by irradiation of visible light and undergoes an energy transfer or a redox path. In the reductive quench path, the (*P.C) absorbs an electron from the electron donor to produce a reduced photo catalyst (P.C$^-$). Which is a good reductant for the one-electron reduction of the substrate (S) or transition metal (M^{n+}). Simultaneously, the photocatalyst is regenerate to the ground state and the reduced species radical anion ($S^{-\cdot}$) or $M^{(n-1)+}$ undergo further reaction. In the oxidative quench pathway, the (*P.C) loses an electron to the electron acceptor to generate an oxidized photo catalyst (P.C$^+$) which is a good oxidant for the one-electron oxidation of the substrate (S) or transition metal (M^{n+}). The photo catalyst is regenerated and oxidative species radical cation ($S^{+\cdot}$) or $M^{(n+1)+}$ could undergo further transformations. Both of these cycles produce $D^{+\bullet}$ and $A^{-\bullet}$ radicals in a single operation through SET to

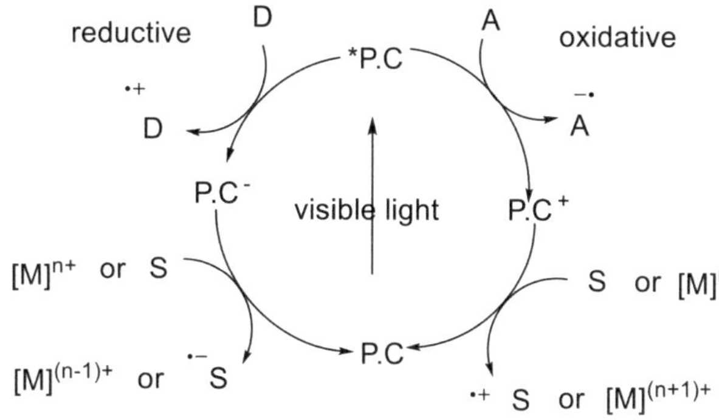

Figure 4.
Redox Photocatalyst via reductive or oxidative pathway.

make the overall process neutral. Here, the reductive path refers to the reduction of the excited photo catalyst (*P.C) where the external electron donor D is oxidized, whereas the oxidative path defines oxidation of the excited photo catalyst (*P.C) with concomitant reduction of the external electron acceptor A (**Figure 4**).

4. Metal-catalyzed C-N bond formations

4.1 Ru-catalyzed C-N bond formations

Several methodologies have been developed for various difunctionalizations of which transition-metal based photocatalytic C-N bond formations is in high demand. Dagousset *et al.* in 2014 reported a metal-catalyzed azido- and aminotrifluoromethylation of alkenes (**2**) from alkene (**1**), azidotrimethylsilane and Umemoto's reagent (**Figure 5**) [9]. The radical-mediated difunctionalization of alkene is promoted under the irradiations of blue LEDs in the presence of a Ru-catalyst. According to the proposed reaction mechanism in the presence of visible light the catalyst $Ru(bpy)_3^{2+}$ form an excites species $[Ru(bpy)_3^{2+}]$ *which generates the CF_3 radical via a single electron transfer (SET) from Umemoto's reagent. The CF_3 radical reacts with the alkene (**1**) providing the radical species with subsequent oxidation to a cation via a SET process from $[Ru(bpy)_3^{3+}]$. Finally, the nucleophilic addition of this β-trifluoromethylated carbocation by $TMSN_3$ or amine afforded the corresponding trifluoro methylated product (**2**).

Yasu *et al.* in 2013 reported a metal-catalyzed facile intermolecular aminotrifluoromethylation of alkenes (**Figure 6**) [10]. This is a highly efficient bifunctional reaction taking place between alkene (**3**), and Umemoto's reagent in MeCN. Here MeCN acts as a *N*-nucleophile, known as an aminative carbocation trap agent (Ritter-type reaction) and Umemoto's reagent serving as the CF_3 source. The reaction takes place via initial SET processes in the presence of blue LEDs through excitation of $[Ru(bpy)_3]^{2+}$ to $^*[Ru(bpy)_3]^{2+}$ which reduce the Umemoto's reagent to produce a CF_3 radical. Then this CF_3 radical attacks alkene (**3**) to give a radical which is further oxidized by $[Ru(bpy)_3]^{3+}$ to form a trifluoromethylated carbocation through another SET process. Finally, the additions of RCN, to the

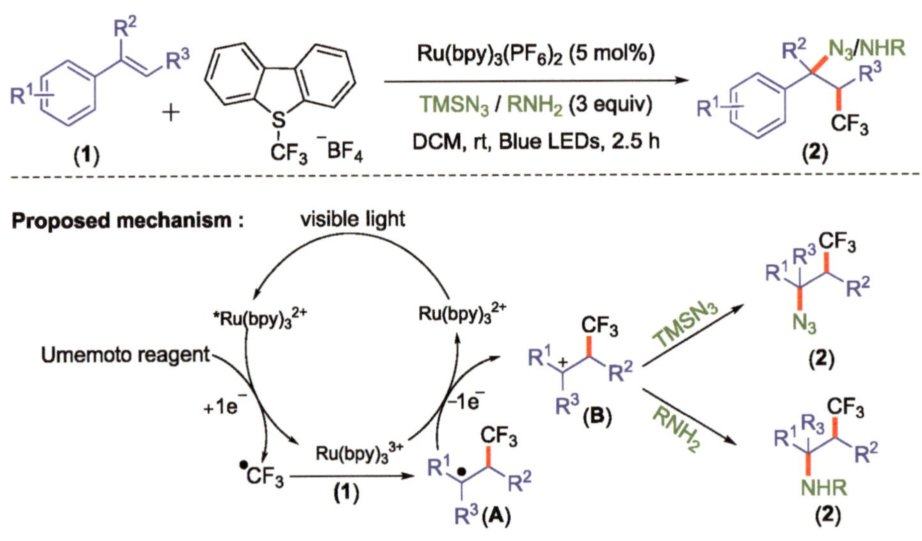

Figure 5.
Synthesis of azido and Aminotrifluoromethylated products.

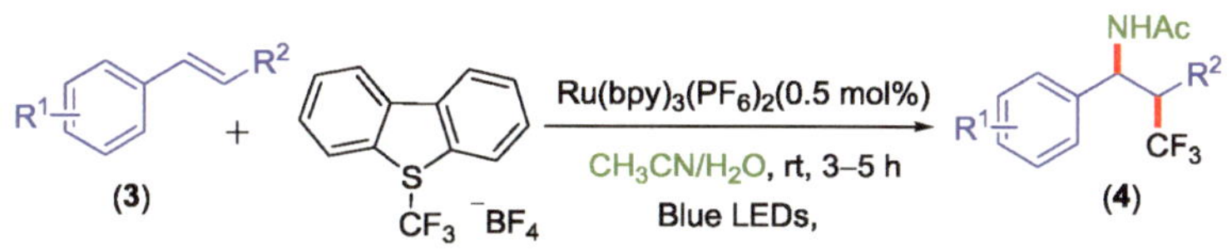

Figure 6.
Synthesis of Aminotrifluoromethylation derivative.

carbocation followed by hydrolysis (Ritter type amination) form the final aminotrifluoromethylated product (**4**).

Carboni *et al.* in 2014 demonstrated an elegant visible-light-mediated method for the synthesis of carbotrifluoromethylation (**6**) of enecarbamates using Togni's reagent as the CF_3 source and NaN_3 as the nucleophile (**Figure 7**) [11]. Herein, the CF_3 radical is generated from the Togni reagent via a reductive photo redox path under a single electron transfer process. Then the addition of enecarbamates (**5**) generates α-amido radical which is rapidly oxidized to an acyliminium cation by a SET process. Finally, nucleophilic additions of NaN_3 affords the product (**6**).

Bearing the importance of C-N bonds in mind Yang *et al.* in 2020 developed an efficient alkylazidation (**8**) using alkene (**7**), sodium azide and heteroareniumsalts as functionalized alkyl reagents for the synthesis of 2-azido-1-(1,4-dihydropyridin-4-yl)-ethane's (**8**). This reaction permits the incorporation of both azido and 1,4-dihydropyridin-4-yl group via difunctionalization of alkenes to construct C-C and C-N bonds in a single operation (**Figure 8**) [12]. The $[Ru(bpy)_3]^{2+}$ specie is excited to $[Ru(bpy)_3]^{2+*}$ by irradiated it with visible-light and undergoes single-electron transfer (SET) with NaN_3 to form an azido radical (N_3). The addition of azido radical across the C=C bond of alkene (**7**) generates an alkyl radical which is followed by the addition of the pyridinyl ring of pyridinium and finally, reductions by the $[Ru(bpy)_3]^+$ species gives a product (**8**) and regenerates the active catalyst.

Yu *et al.* in 2016 disclose a Ru-catalyzed visible-light-mediated synthesis of azotrifluoromethylation (**10**) in the presence of alkenes (**9**) with aryldiazonium salts and sodium trifluoromethanesulfinate (**Figure 9**) [13]. This reaction is successful for unactivated alkene. Both electron-donating and electron-withdrawing groups of alkene and aryldiazonium salts give their product (**10**) in good yields. These trifluoromethylated azo products are useful building blocks for many heterocycles and nitrogen-containing compounds. As per the suggested mechanism in (**Figure 9**) in the presence of blue light, photoexcitation of the photocatalyst

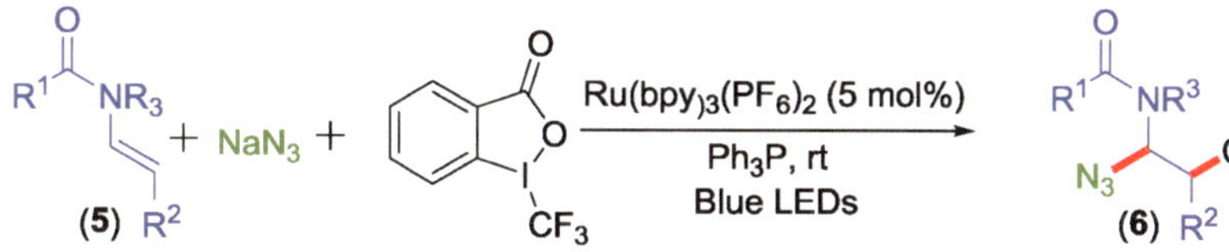

Figure 7.
Synthesis of Carbotrifluoromethylation derivative.

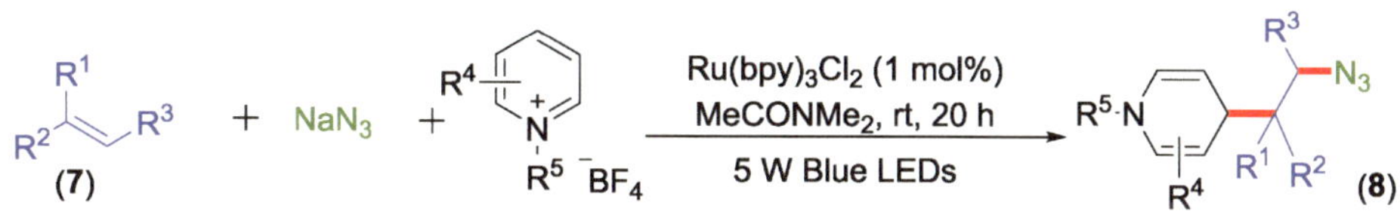

Figure 8.
Synthesis of 2-Azido-1-(1,4-dihydropyridin-4-yl)-ethanes.

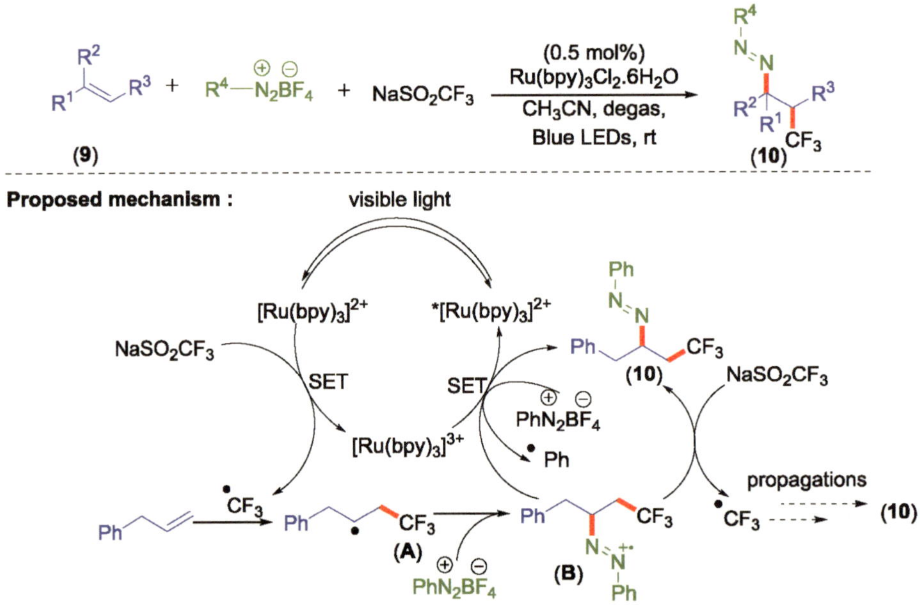

Figure 9.
Synthesis of Trifluoromethylated azo compounds.

$[Ru(bpy)_3]^{2+}$ generates the excited state $^*[Ru(bpy)_3]^{2+}$ species. This active species is transferred into the oxidizing photocatalyst $[Ru(bpy)_3]^{3+}$ via SET oxidation by the phenyldiazonium salt. This active species serves as a strong oxidant to oxidize Langlois' reagent to produce CF_3 radical upon removal of SO_2 and returning the photocatalyst to its ground state. At this time, the CF_3 radical undergoes addition to the alkene (**9**) to generate the radical intermediate (A), which is easily trapped by the aryldiazonium salt to give the radical cation intermediate (B). Similarly, another SET reduction of intermediate (B) by the reducing $^*[Ru(bpy)_3]^{2+}$ species gives to the desired product (**10**).

Since vicinal diamine are found in many pharmaceuticals and various biologically active compounds hence the development of newer methodologies is deemed worthy. Considering their biological importance and ongoing demand Govaerts *et al.* in 2020 demonstrated a Ru-catalyzed diamination of alkene (**11**) in the presence of blue LEDs (**Figure 10**) [14]. This methodology exploits the generation of aminium radicals from the in situ generated *N*-chloroamines and their capability to react with alkenes via anti-Markovnikov addition.

According to the depicted mechanism (**Figure 10**) initially, chlorination of an alkylamine with NCS occurs followed by the addition of a strong Brønsted acid generates a highly activated *N*-chloroammonium intermediate (A) which upon SET via photoexcited state of $Ru(bpy)_3^{2+}$ creates the intermediate (B). This amminium radical intermediate (B) adds to the olefin (**11**) to give anti-Markovnikov intermediate (C). Simultaneously the β-ammonium radical intermediate (C) reacts with intermediate (A) restoring the intermediate (B) and provide the protonated β-chloroamine intermediate (D) which gives intermediate (E) followed by ring-opening by a second alkyl amine (e.g., Et_2NH) to give the product (**12**).

Considering the importance of acyl amide, Hari *et al.* in 2014 established a photochemical method (**Figure 11**) for the synthesis of functionalized amide (**14**) using diazonium salt as the cheap and environment-friendly arylation partner and alkene (**13**). [15] The photo Meerwein arylation reaction is applied only for the formation of aryl–alkene coupling products. As suggested in the mechanism,

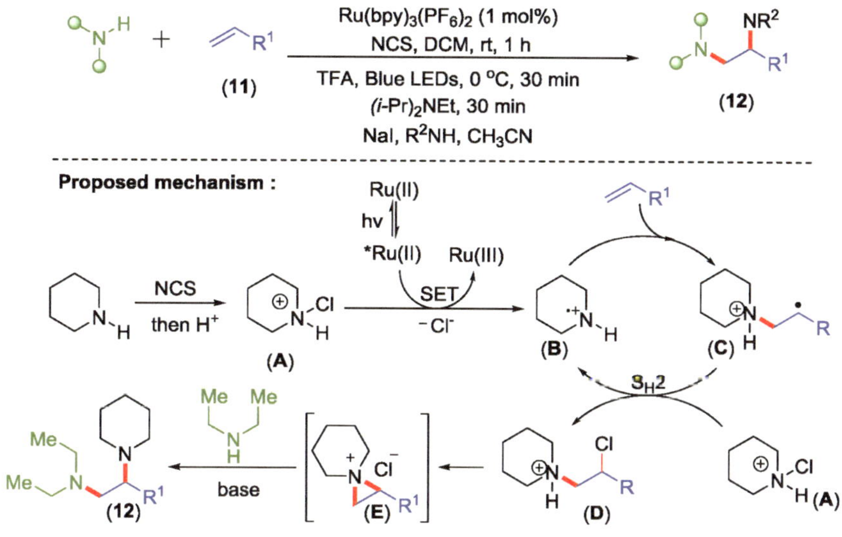

Figure 10.
Synthesis of 1, 2-Diamination product.

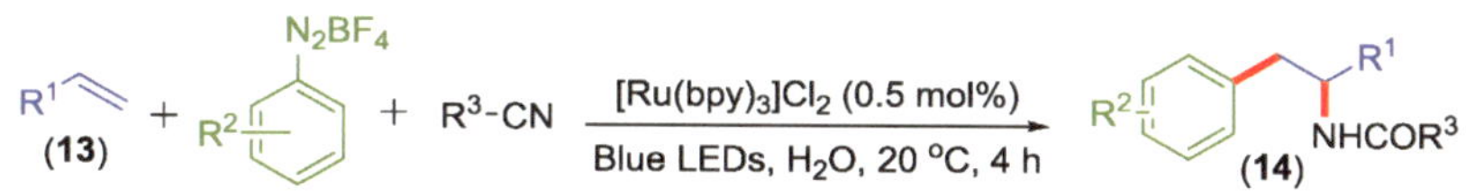

Figure 11.
Synthesis of amino-Arylation product.

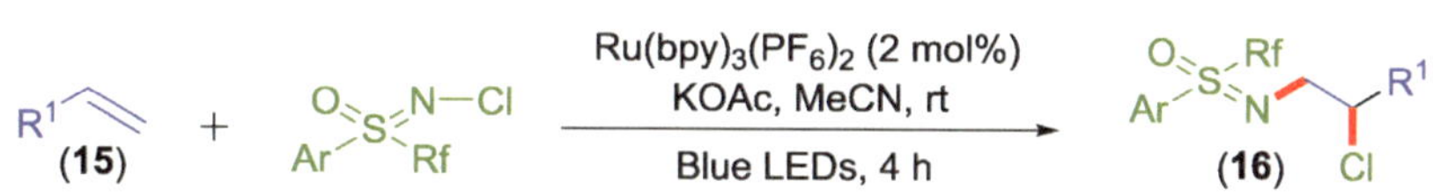

Figure 12.
Synthesis of N-Chloro S-Fluoroalkyl Sulfoximines.

initially an aryl radical is formed via a single-electron transfer (SET) from the excited state of the photocatalyst $[Ru(bpy)_3]^{2+*}$ to a diazonium salt. Then the addition of aryl radical to alkene (**13**) generates another radical intermediate which undergoes oxidation to provide a carbenium species. Finally, the attack of a nitrile (R_3CN) to the carbenium species followed by hydrolysis gives the amino-arylated product (**14**).

Considering the biological importance of sulfoximines containing compounds, Prieto *et al.* in 2019 demonstrated a method for the formation of *N*-chloro S-fluoroalkyl sulfoximines (**16**) from alkene (**15**) and sulfoximine through an atom transfer radical addition (ATRA) mechanism. A broad reaction scope was demonstrated, and various functionalised sulfoximines were well tolerated in the present protocol (**Figure 12**) [16]. Herein the photoexcited catalyst reacts with sulfoximine by SET reduction to give the sulfoximidoyl radical which is then followed by reaction with the alkene providing the alkyl radical. At this time, two different paths are possible, one is via the radical-chain path and another the catalytic path. In the radical-chain pathway the alkyl radical abstracts a chlorine atom from sulfoximine to give the compound (**16**) and generate a new sulfoximidoyl radical. In the catalytic pathway, the intermediate alkyl radical undergo oxidation by the oxidized form of PC into a cationic species and restore the photocatalyst. Finally, the addition of chlorine atom to the cationic species afforded the compound (**16**).

Ouyang *et al*. in 2018 established an elegant method in which a photo-induced three-component reaction of styrenes (**17**) with alkyl *N*-hydroxyphthalimide (NHP) esters and amine leads to 1,2-alkylamine (**18**) (**Figure 13**) [17]. In this reaction, the alkyl NHP esters act as an alkylating agent to give 1,2-alkyl amine products from their respective alkenes. The plausible mechanism involves a visible-light excitation of the photo redox catalyst thereby decomposing the alkyl NHP ester to an alkyl radical, CO_2, and phthalimide anion. Then the addition of alkyl radical across the C=C bond of arylalkene (**17**) generates another alkyl radical which upon single electron transfer through oxidation of the $[Ru(bpy)_3]^{3+}$ species provide the alkyl cation. Finally, the nucleophilic attack of amine to the cationic species delivers the final product (**18**).

4.2 Ir-catalyzed C-N bond formations

Miyazawa *et al*. in 2015 demonstrated a regiospecific synthesis of aminohydroxylation (**20**) from alkenes (**19**) by photo redox catalysis (**Figure 14**) [18]. Here *N*-protected 1-aminopyridinium salt is the key compound that provides an amidyl radical precursor in the presence of Ir-photocatalyst. The reaction proceeds via an Ir-catalyzed radical-mediated path in the presence of acetone and water under the irradiations of blue LEDs providing difunctionalized alkenes. The proposed mechanism is shown in (**Figure 14**). In the presence of visible light, the photocatalyst Ir^{III} is excited to $^*Ir^{III}$, which undergoes single electron transfer (SET) to an aminopyridinium to provide a stabilized radical (A) and a highly oxidizable Ir species Ir^{IV}. The generated amidyl radical from intermediate (A) reacts with alkene (**19**) in a regiospecific manner to give a radical intermediate (B). Then Ir^{III} is oxidized to form an Ir^{IV} species and afford β-amino carbocation intermediate (C)

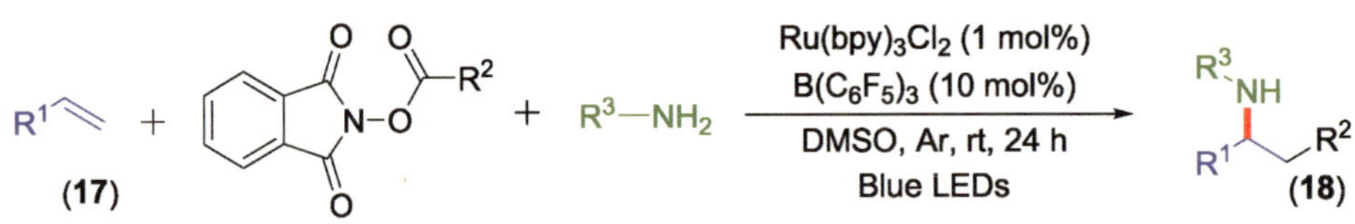

Figure 13.
Synthesis of 1, 2-alkyl amination product.

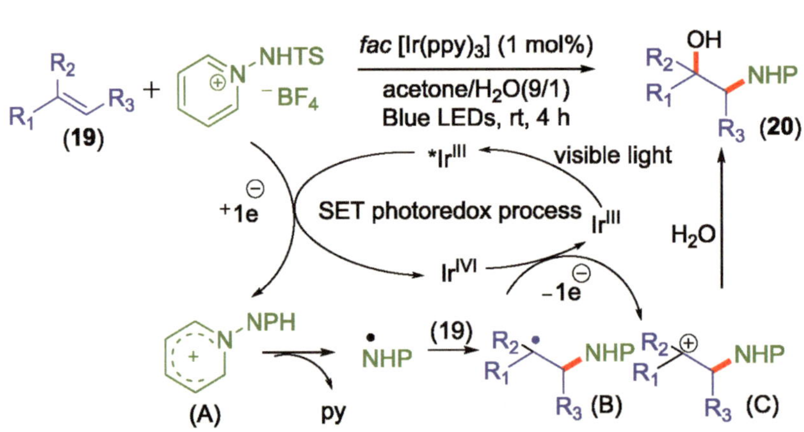

Figure 14.
Synthesis of 1,2-Aminoalcohol.

and regenerate the Ir photocatalyst to its ground state Ir^{III}. Finally, the nucleophilic attack of H_2O to the carbocation intermediate (C) produce the product 1,2-aminoalcohol (**20**).

Xu and Cai in 2019 reported a metal-catalyzed visible-light-mediated difunctionalization of alkene (**21**) where $BrCF_2CO_2Et$ and amines are the coupling partner (**Figure 15**) [19]. The present strategy is equally successful for electron-poor, electron-rich, and internal alkenes. The Csp^3–Csp^3 and Csp^3–N bonds are simultaneously formed under mild conditions. According to the proposed mechanism in (**Figure 15**) initially, in the presence of visible light, *fac*-$Ir^{III}(ppy)_3$ is excited to *fac*-$Ir^{III}(ppy)_3^*$, which then reacts with $BrCF_2CO_2Et$ by the SET pathway to generate the ethyl difluoroacetate radical (A) and the oxidized photocatalyst *fac*- $Ir^{IV}(ppy)_3$. Then, the selective addition of the ethyl difluoroacetate radical (A) to the alkene (**21**) generate a benzyl radical intermediate (B). This intermediate (B) undergo single-electron oxidation via the cooperative effects of the active Fe (III) species, to form the carbocation intermediate (C). Finally, the attack of an arylamine to the intermediate (C) is followed by base-mediated deprotonation to generate the difluoroalkylamination product (**22**).

Wu *et al.* in 2019 demonstrated a metal-catalyzed synthesis of β-arylsulfonyl (diarylphosphinoyl)-α,α-diarylethyl-amines (24) from readily available 1,1-diarylethylenes (**23**), arylazides, and arylsulfinic acids (**Figure 16**) [20]. This Ir-catalyzed reaction takes place in the presence of blue LEDs and under the nitrogen atmosphere leading to difunctionalizations of alkenes. As per the proposed mechanism (**Figure 16**) by the irradiations of blue light, the catalyst $[Ir(mppy)_3]^{3+}$ is excited to $[Ir(mppy)_3]^{3+*}$ through energy transfer. Then, arylsulfinic acid and $[Ir(mppy)_3]^{3+*}$ participate in a SET process to generate $[Ir(mppy)_3]^{2+}$ and arylsulfonyl radical (A). Then the addition of alkene (**23**) to the intermediate (A) form the α α-diarylalkyl radical (B). Simultaneously, $[Ir(mppy)_3]^{3+*}$ transfers its energy to the arylazide resulting in loss of N_2 and the constructions of triplet nitrene intermediate (C). After that via a SET process and protonation, the intermediate (C) is transferred to a nitrogen radical intermediate (D) which then adds to the persistent radical intermediate (B) to give the β-arylsulfonyl(diarylphosphinoyl)-α,α-diarylethyl-amines (**24**).

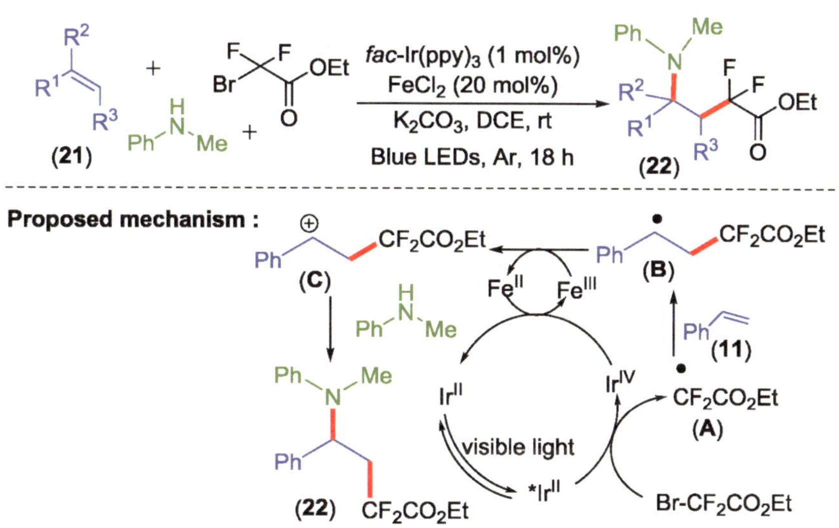

Figure 15.
Synthesis of Difluoroalkylamination product.

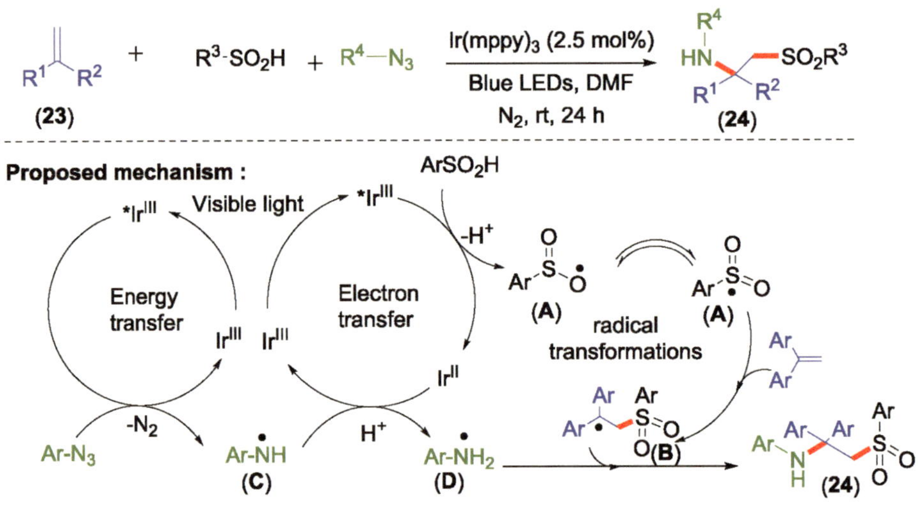

Figure 16.
Synthesis of β-Arylsulfonyl-α, α-diarylethylamines.

Chen *et al.* in 2019 described an Ir-catalyzed visible-light-mediated azidoarylation of alkenes in the presence of pyridines and $TMSN_3$ (**Figure 17**) [21]. These reactions take place in the presence of *tert*-butanol and irradiations of 90 W blue LEDs. Electron-withdrawing and electron-donating group of alkene and cyanopyridine react smoothly to give the product (**26**). According to the proposed mechanism, irradiation of $Ir(ppy)_2(dtbbpy)PF_6$ produce an excited state Ir* which would capture a single-electron from azide to generate the azido radical intermediate (A) and reducing photocatalyst Ir^{II}. Then the addition of electrophilic azido radical intermediate (A) to the alkene (**25**), produce a benzylic radical intermediate (B). A single-electron reduction between Ir^{II} and cyanopyridine generate the pyridyl radical anion intermediate (C) and regenerate the ground-state Ir^{III} catalyst. Simultaneously, a radical–radical coupling between the transient benzylic radical intermediate (B) and the pyridyl radical anion intermediate (C) afford the intermediate (D), which can undergo elimination of a CN anion give the product (**26**).

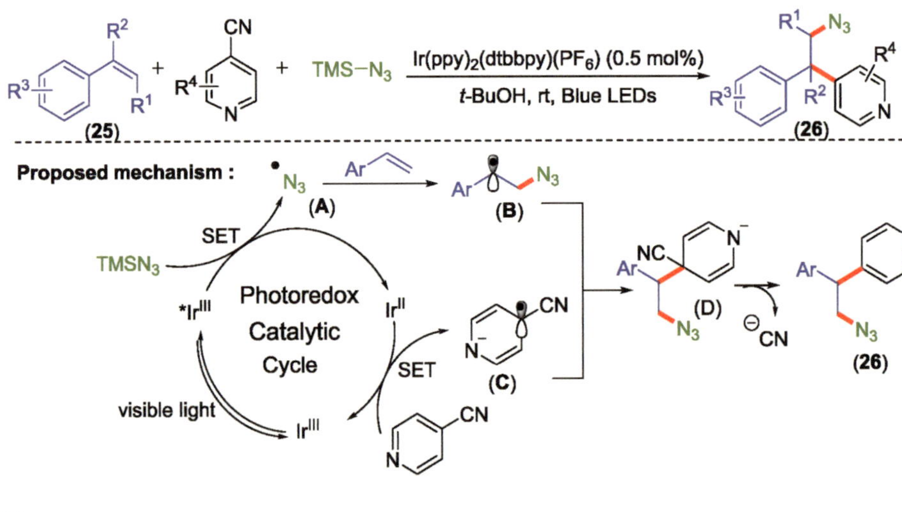

Figure 17.
Synthesis of β–Azidopyridines.

Recently, Guo *et al*. in 2021 demonstrated a photocatalytic 1,2-diamination of 1,3- dienes (**27**) in the presence of *N*-aminopyridinium and TMSNCS to affords 1,2-aminoisothiocyanation products (**28**) in high chemo- and regio-selective manner with broad substrate scope and good functional group tolerance (**Figure 18**) [22]. According to the proposed mechanism (**Figure 18**) the visible light excites the Ir-catalyst which reduced *N*-aminopyridinium salt to produce a radical intermediate (A) and underwent dissociations to generate a nitrogen-cantered radical (B) and 2,4,6-collidine. Regioselective addition of intermediate (B) to 1,3-diene (**27**) afford an allylic intermediate (C). Eventually, oxidation of the intermediate (**C**) by Ir^{IV} afforded a carbocation intermediate (D) with concurrent regeneration of Ir^{III} species. Nucleophilic addition of TMSNCS to the intermediate (D) generated a mixture of 1,2-aminoisothiocyanation (**28**) and 1,2-aminothiocyanation products.

An elegant method for the synthesis of β-sulfonyl amides (**30**) is reported by Zong *et al.* in 2019 through an acid promoted photochemical reaction of styrenes (**29**), aryldiazonium tetrafluoroborates, sulfur dioxide, nitriles, and water (**Figure 19**) [23]. This visible-light-mediated vicinal aminosulfonylation of an alkene with the insertion of SO_2 giving rise to β-sulfonyl amides (**30**) with high efficiency and excellent chemoselectivity, in moderate to good yields.

As depicted in **Figure 19** the plausible mechanism involves the interaction of aryldiazonium tetrafluoroborate with DABCO·$(SO_2)_2$ to generate aryl radical, sulfur dioxide, nitrogen, and DABCO radical cation. Then the aryl radical is captured by sulfur dioxide to generate an aryl sulfonyl radical intermediate (A) which subsequently attacks the alkene (**29**) to furnish a C-centered radical (B). Next with the help of photocatalyst via oxidative SET of C-central radical intermediate (B), provide a cation intermediate (C). Then the attack of nitrile to cation intermediate (C), generated another cation intermediate (D). Water acts as a nucleophile in the presence of a Lewis acid to attack the cation intermediate (D), leading to compound (E). The subsequent isomerization gives β-sulfonyl amide (**30**).

Ge *et al*. in 2020 developed a metal-catalyzed three-component reaction of alkene (**31**), using a selenium ylides-based trifluoromethylation reagent, and nucleophiles such as azide, amine, alcohol, water via a radical process to

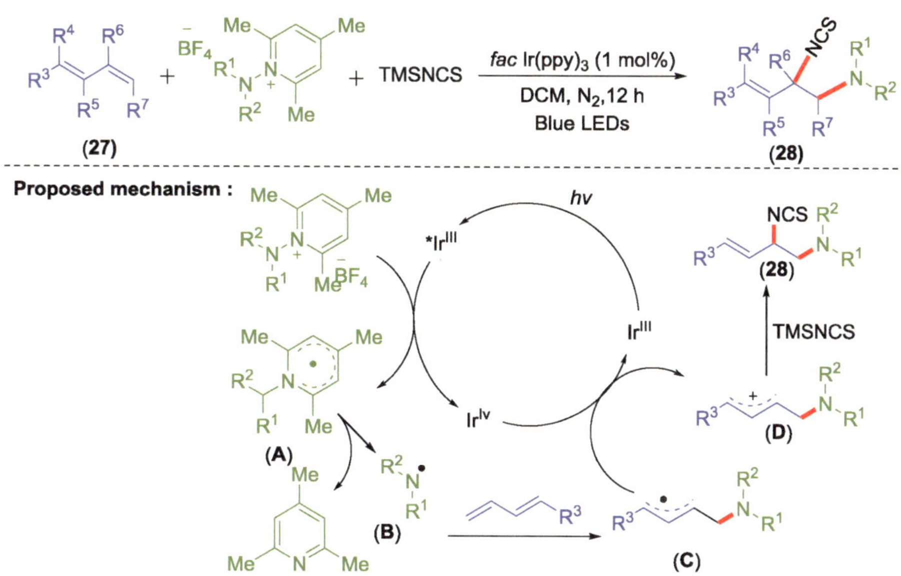

Figure 18.
Synthesis of 1,2-Diamination product.

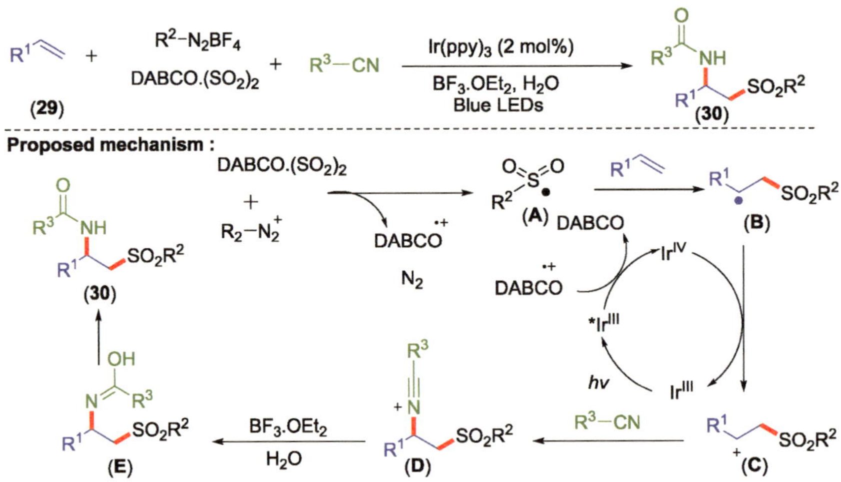

Figure 19.
Synthesis of β-sulfonyl amides.

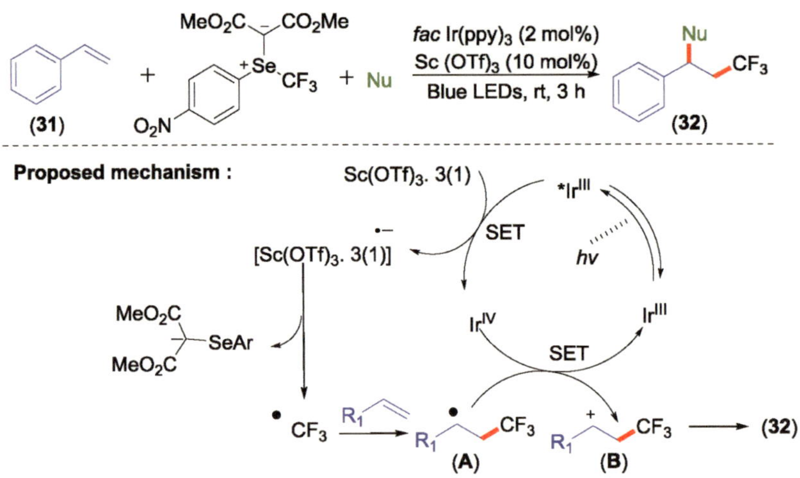

Figure 20.
Synthesis of Trifluoromethylative amination product.

trifluoromethylative amination product (**32**) under mild conditions (**Figure 20**) [24]. The trifluoromethylation reagent act as a trifluoromethyl radical source. The process takes place in the presence of a Lewis acid scandium(III) trifluoromethanesulfonate $Sc(OTf)_3$ and a photoredox catalyst [fac-Ir(ppy)$_3$]. As per the proposed mechanism (**Figure 20**) initially, irradiation of blue light excited the photocatalyst [fac- Ir^{III}(ppy)$_3$] which then transfers one electron to [$Sc(OTf)_3$•3(1)] (obtained by mixing selenium ylide-based trifluoromethylating reagent with $Sc(OTf)_3$ via a single electron transfer(SET) process). This radical anion $[Sc(OTf)_3\bullet 3(1)]^{\bullet -}$ is unstable and undergoes homolytic cleavage of Se – C(CF_3) bond to generate the CF_3• radical. This CF_3• radical addition to the styrene (**29**) gives a trifluoromethylated benzylic radical intermediate (A), which is oxidized by [Ir^{IV}(ppy)$_3$] to a benzylic cation intermediate (B). Finally, the nucleophilic attack at the benzylic cation gives trifluoromethylated product (**30**).

Qin *et al*. in 2017 demonstrated an Ir-catalyzed protocol for the synthesis of α-amino ketones and diaminations product (**34** and **34′**) from activated olefins (**33**)

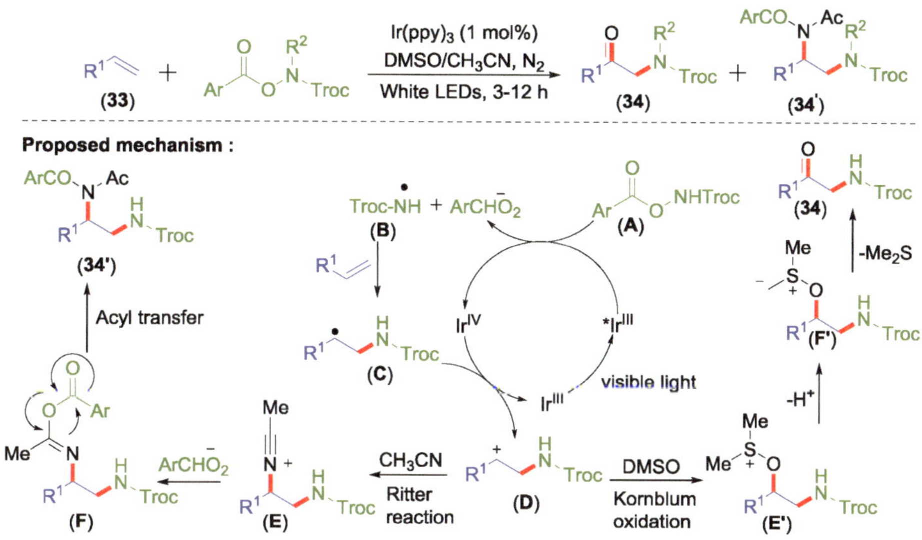

Figure 21.
Synthesis of 1,2-Diamides and α-amino ketones.

(**Figure 21**) [25]. Here o-acyl hydroxylamines are the key reacting partner for difunctionalization. Here solvent also play a pivotal role as different solvent gave different products. A plausible mechanism for the Ir-catalyzed radical diamination and α-amino ketone of active olefins is shown in (**Figure 21**). In the presence of visible-light the exited Ir-catalyst induce the reductive cleavage of reactant (A) to generate a radical (B) and a carboxylate anion. The *N*-centered radical (B) adds to the styrene (**33**) to produce an alkyl radical intermediate (C) with the regeneration of photocatalyst Ir^{III} via oxidation to give a carbocation intermediate (D). In CH_3CN, (D) is trapped by the solvent to give a nitrilium intermediate (E) through a Ritter-type process. Then attack by the intermediate (E) followed by an acyl migration afford the diamidated product (**34′**). When DMSO is used as the solvent, intermediate (D) can also be trapped by the solvent to provide an alkoxysulfonium intermediate (E’), which undergoes a Kornblum oxidation to afford the α-amino ketone (**34**).

Qin *et al*. in 2015 demonstrated an Ir-catalyzed visible-light-mediated synthesis of chlroamines (**36**) from activated olefins (**35**) (**Figure 22**) [26]. Here *N*-chlorosulfonamides served both as nitrogen and chlorine source. This methodology provides regioselective, efficient, and atom-economic method for the preparation of vicinal halo amines. The reaction goes via the generation of a nitrogen-centered radical from *N*-chlorosulfonamide by oxidative quenching of the Ir-catalyzed which is excited in the presence of blue LEDs. This nitrogen centered radical then adds to the olefin (**35**) to produce an alkyl radical which is further oxidized to a carbocation intermediate with the regeneration of Ir^{III}. Finally, the addition of chloride anion to the carbocation gives chloraminated product (**36**).

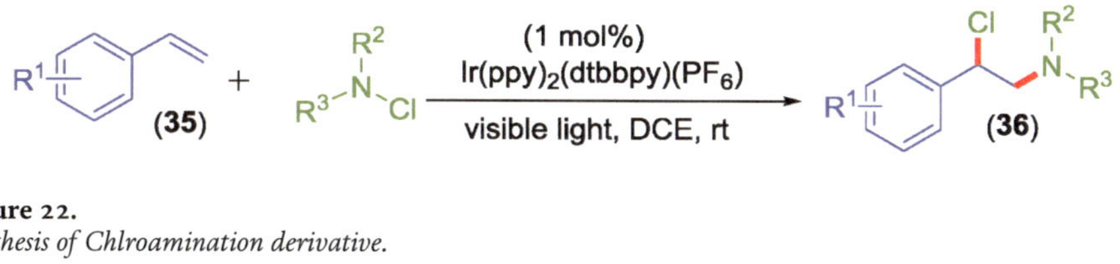

Figure 22.
Synthesis of Chlroamination derivative.

4.3 Cu and Pd-catalyzed C-N bond formations

Wu *et al.* in 2019 depicted (**Figure 23**) a visible-light-mediated Cu-catalyzed difunctionalization of alkene (**37**) to give azidation product (**38**) [27]. Here the azidobenziodoxole acts as an azidating agent in the presence of acetonitrile and $[Cu(dap)_2]PF_6$ complex as the photocatalyst. While the reactions produced three types of difunctionalized products, which correspond to reaction patterns of amido-azidation, diazidation and benzoyloxy-azidation. The electronic factor of the aryl group attached to the alkene play a vital role in determining the reaction outcome. When the aryl group is rich in electron, the reaction afforded benzoyloxy-azidation product and highly electron-deficient vinyl arenes, generated diazidation products in moderate yields. When the aryl group is electron-deficient or moderately electron-rich, give predominantly amido-azidation product. Based on the proposed mechanism the reaction is initiated via single electron transfer (SET) between IBA-N_3 and $[Cu(dap)_2]^{+*}$, which provided an azidyl radical and $[Cu(dap)_2]^{2+}$. The azide radical then attacks the alkene to produce a radical which would couple with the CH_3CN-$[Cu]^{2+}$ complex followed by reductive elimination to give Ritter-type intermediate. The latter is readily captured by the *o*-iodo benzoyloxylate anion and further rearrangement formed product (**38**). In another path, the radical generated from alkene is oxidized to the corresponding carbocation by IBA-N_3. Then addition with *o*-iodo benzoyloxylate anion afforded the oxy-azidation products (**38′**). When the vinyl group is attached to a strong electron-withdrawing group intermediate generated from alkene abstracts an azidyl group from IBA-N_3 to give the diazidation product (**38″**).

Similarly, Fumagalli *et al.* in 2015 demonstrated a method for the synthesis of azidation derivatives (**40** and **40′**) from activate alkenes (**39**) (**Figure 24**) [28]. The reaction is light-switchable, in the presence of light gives methoxyazidated products and in the absence of light, diazidation product is obtained. This methodology uses sustainable and cheap copper-based photocatalyst, to enable electron transfer under mild reaction conditions, thus affecting the formation of double C-N bond in dark, and C-N/C-O formation in the presence of light.

Hossain *et al.* in 2018 developed a visible-light-photocatalytic strategy for the synthesis of azido ketones (**42**) from vinyl arenes (**41**) and $TMSN_3$ (**Figure 25**) [29]. The reactions proceed via step-economic fashion under an aerobic condition without additional oxidants. As per the mechanism (**Figure 25**) initially in the presence of light $[Cu(dap)_2]Cl$ gets excited and form an excited state intermediate (A) via oxidation with dioxygen with the release of an equivalent of ligand (dap).

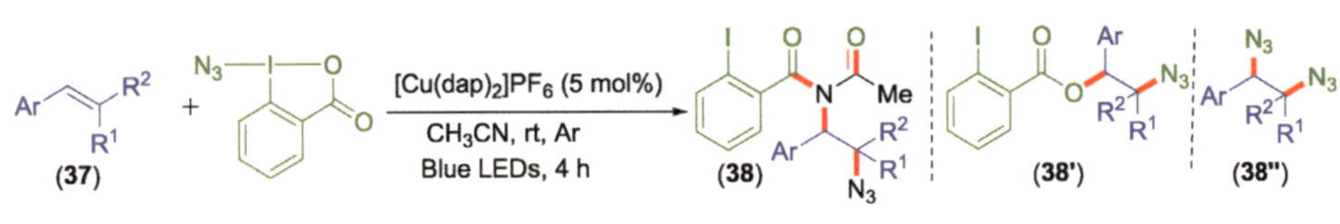

Figure 23.
Synthesis of Azidation derivative.

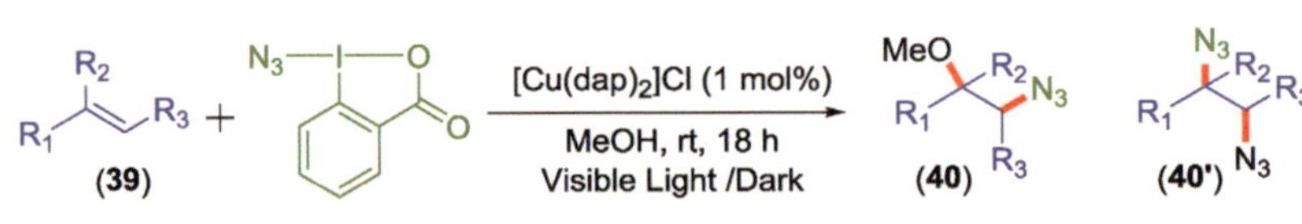

Figure 24.
Synthesis of Azidation derivative.

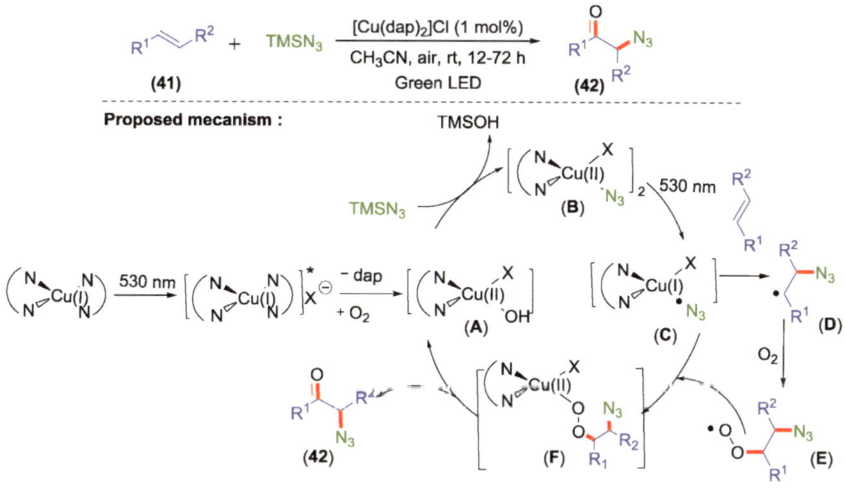

Figure 25.
Synthesis of azido ketones.

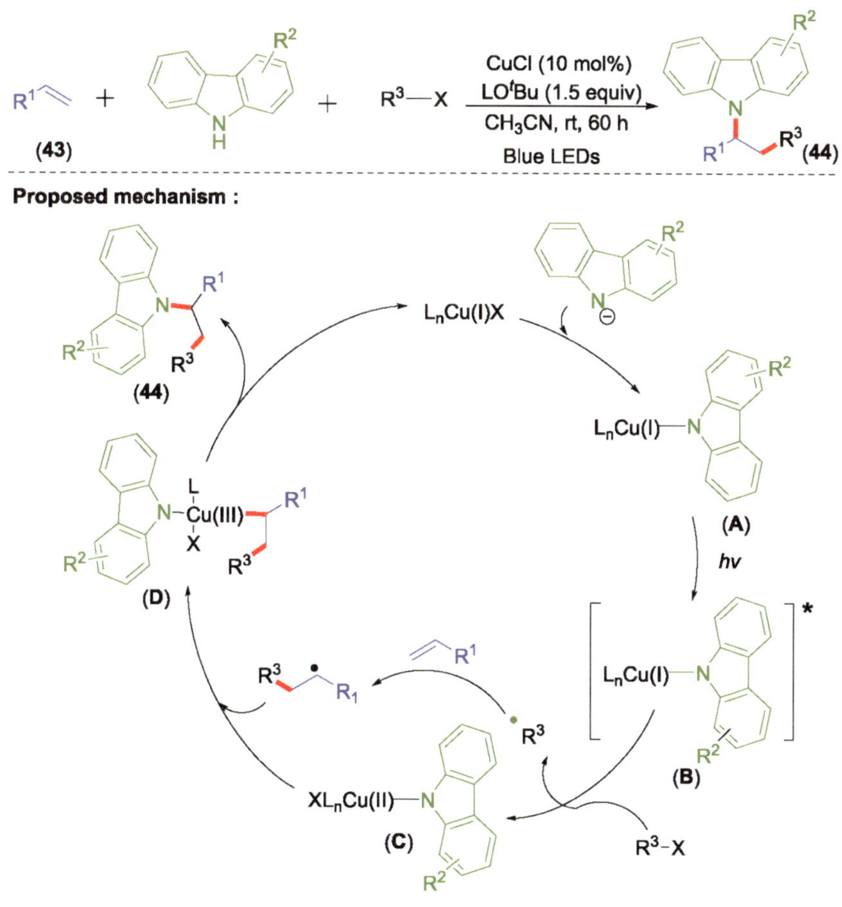

Figure 26.
Synthesis of Carboaminations product.

Then the formation of an intermediate (B) rapidly occurs upon mixing of [Cu(dap) Cl_2] with $TMSN_3$. The homolytic dissociation of intermediate (B) generates another intermediate (C) and an azide radical. After that the azide radical attacks the alkene to form a stabilized radical intermediate (D), which further reacts with oxygen to form intermediate (E). Then this radical intermediate binds to intermediate(C) to form a Cu^{II} species intermediate (F). Finally, the release of intermediate (A) gives the ketoazide product (**42**).

Xiong *et al*. in 2019 reported a Cu-catalyzed visible-light-mediated synthesis of aminoalkylated derivative (**44**) in the presence of alkene (**43**), alkyl iodides, and carbazole (**Figure 26**) [30]. According to the proposed mechanism (**Figure 26**) initially, L_nCuCl undergoes ligand exchange with the nucleophile to deliver $L_nCu(I)Nu$ (A). Then it goes to the excited-state adduct intermediate (B) in the presence of LEDs. The excited-state intermediate (B) would then involve an electron transfer with the halide to generate a radical and intermediate (C). Next, the radical attack to the alkene produces an internal radical and intermediate (D). Finally, the product (**44**) undergo reductive elimination and regenerate the Cu(I) catalyst.

Cheung *et al*. in 2020 disclose a Pd-catalyzed visible-light-mediated synthesis of aminoalkylations derivative (**46**) via 1,2-carbofunctionalization of conjugated dienes (**45**) using alkyl iodides and amines as the coupling partners (**Figure 27**) [31]. This methodology is subsequently utilizing for the late-stage derivatization of complex molecules which is useful in drugs discovery. The multi-component reaction uses readily available reaction partners with broad substrate scope and does not require any exogenous photosensitizers or external oxidants.

The proposed mechanism is shown in (**Figure 27**), at first, the photoexcited $LnPd^0$ undergo single electron transfer (SET) with alkyl iodide to generate a hybrid alkyl palladium radical intermediate (A), which attack at the terminal position of diene (**45**) to give the radical intermediate (B). This intermediate exists in equilibrium with the π-allyl complex intermediate (C). A subsequent nucleophilic attack of the amine forms the carbofunctionalization product (**46**) and regenerates the palladium catalyst.

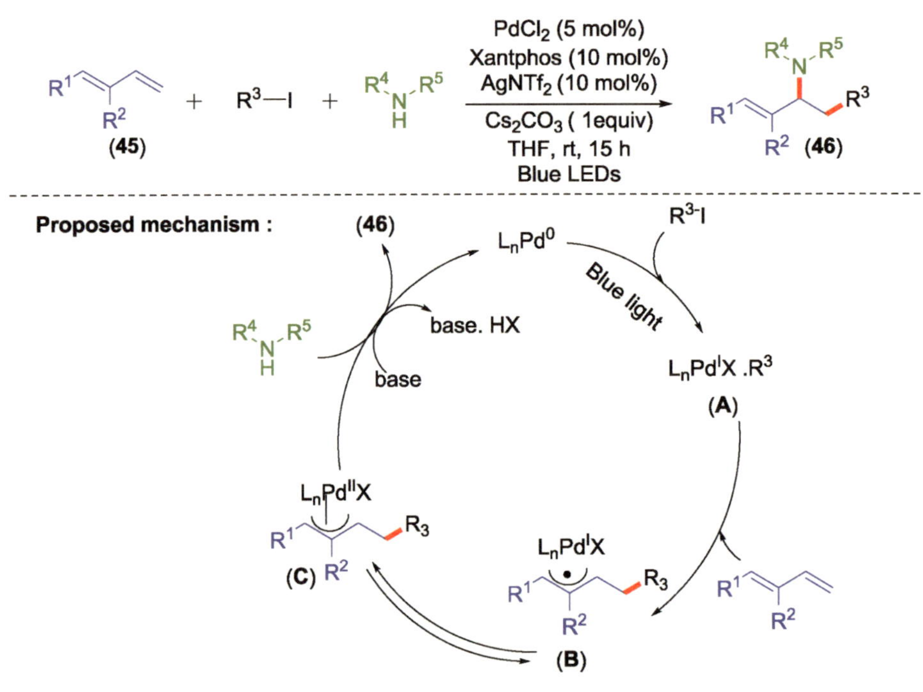

Figure 27.
Synthesis of Aminoalkylations derivative.

5. Organic dye-catalyzed C-N bond formations

5.1 Rose Bengal and 9-Fluorenone catalyzed C-N bond formation

Wei *et al*. in 2018 reported a new and facile visible-light-mediated synthesis of α-azido ketones (**48**) via oxyazidation of alkenes (**47**) with $TMSN_3$ in the air at room temperature (**Figure 28**) [32]. Rose Bengal is a metal-free photocatalyst, used for the synthesis of α-azido ketones. This difunctionalized products are easily and efficiently obtained in moderate to excellent yields via the formation of C-N and C=O bonds. The proposed mechanism is shown in (**Figure 28**). At first, visible-light irradiation of Rose Bengal generated the excited RB*. Subsequently, a single electron transfer (SET) process takes place between $TMSN_3$ and RB* to produce an azido radical and $RB^{\bullet-}$ radical anions. Then, the ground state Rose Bengal and $O_2^{\bullet-}$ is formed through the oxidation of $RB^{\bullet-}$ by molecular oxygen (air). Furthermore, the attack of azido radical to alkene (**47**) gives the alkyl radical intermediate (A). Next, the interaction between radical intermediate (A) with $O_2^{\bullet-}$ and H_2O afforded hydroperoxide intermediate (B). The oxidation of $(PhSe)_2$ by hydroperoxide (B) yielded organoselenium intermediate (C) and PhSeOH. Finally, the additions of PhSeOH with reactive intermediate (C) produced the product (**48**) and regenerated $(PhSe)_2$ and elimination of water takes place.

Wang *et al*. in 2019 depicted a visible-light-induce intermolecular azido-hydrazination method for the synthesis of β-azido alkyl hydrazines (**50**) from unactivated alkenes (**49**) (**Figure 29**) [33]. This transformation occurs via metal-free and redox neutral conditions and applies to a wide range of alkenes. The β-azido alkyl hydrazines are used for the preparations of many valuable synthetic building blocks. As per mechanism (**Figure 29**) initially, photo-excited fluorenone generates a N_3 radical via the catalytic oxidation of azide source with the formation of a reductive ketyl radical species (B). Then, the azide radical attacks the alkene to

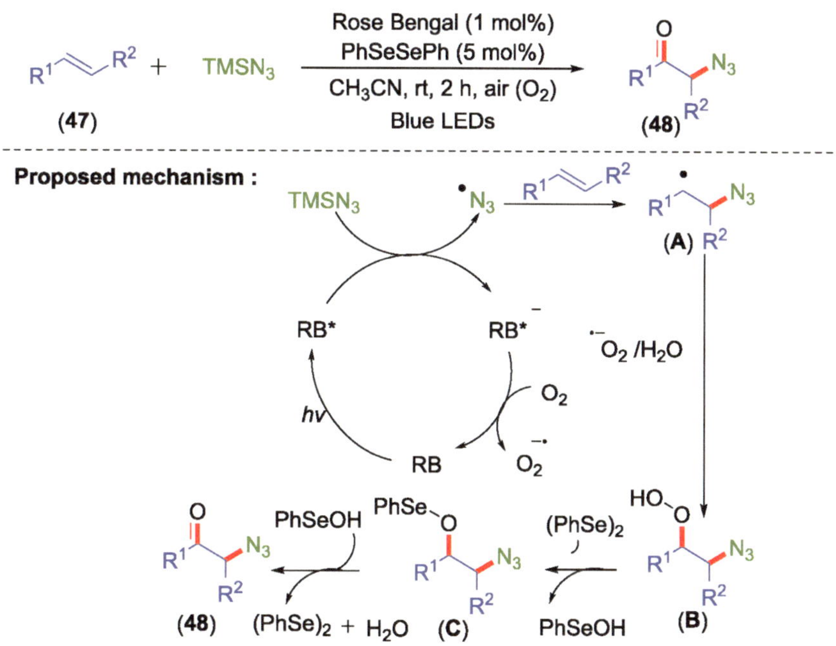

Figure 28.
Synthesis of azido ketones.

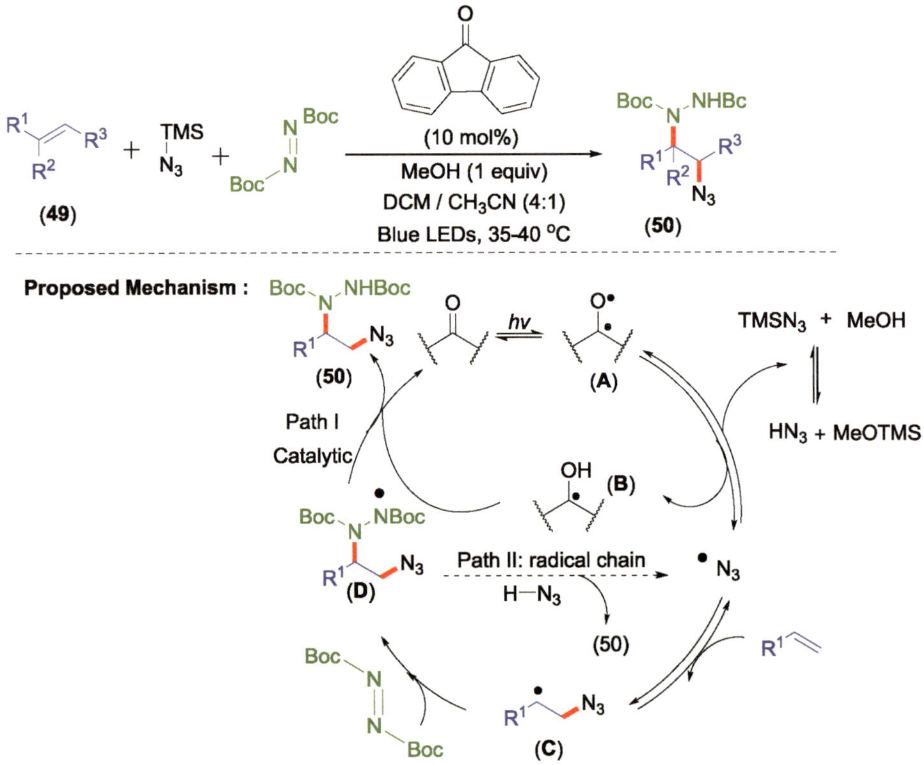

Figure 29.
Synthesis of β-azido alkyl Hydrazines.

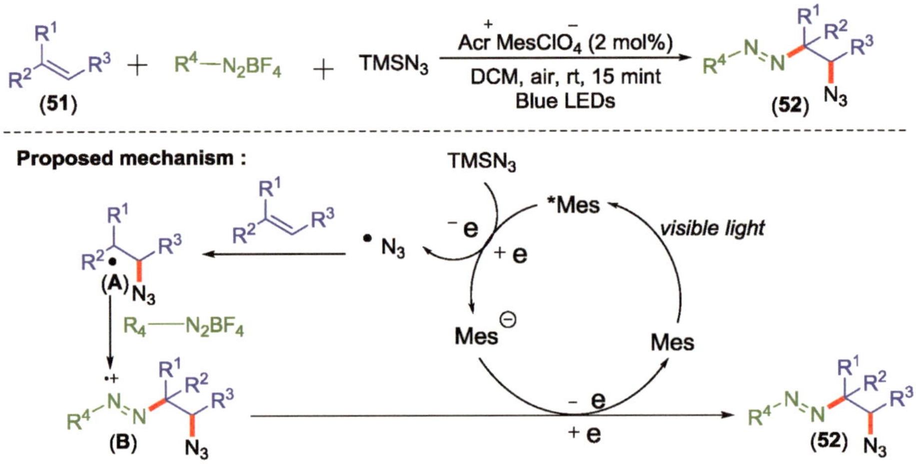

Figure 30.
Synthesis of unsymmetrical azo compounds.

give the alkyl radical intermediate (C). After that, the alkyl radical (C) is trapped by azodicarboxylate giving an *N*-centered radical intermediate (D) which upon the abstraction of a proton give the final product (**50**) (path I). On the other side, the intermediate (D) is reduced by H-N_3 to regenerate an azide radical and the cycles continue (path II).

5.2 Mes-catalyzed C-N bond formation

Shen *et al.* in 2021 reported a photo-induced multi-component cascade reaction in the presence of aryldiazonium salts with unactivated alkenes (**51**) and trimethylsilyl azide ($TMSN_3$) under oxidant-free conditions (**Figure 30**) [34]. This protocol provides a new synthetic method for unsymmetrical azo compounds and applies to different aryldiazonium salts and alkenes. According to the proposed mechanism (**Figure 30**) initially, in the presence of visible light, the photocatalyst 'Mes' undergoes an excited state (*Mes) and take parts in a single electron transfer (SET) process with $TMSN_3$ to generate the azido radical and 'Mes'$^{-}$ radical anion. Subsequently, the azido radical attacked the alkene (**51**) to produce an alkyl radical intermediate (A), which is trapped by the aryldiazonium salt to generate a radical cation intermediate (B). Finally, another SET process between the radical cation (B) and the 'Mes'$^{-}$ radical anion provide the product (**52**) with simultaneous regeneration of the photocatalyst 'Mes'.

Yang and Lu in 2017 established a suitable method for the formation of hydroxyazidation derivative (**54**) from the reaction of alkenes (**53**) under visible light photo redox catalysis (**Figure 31**) [35]. The important features of the reaction are low catalyst loading, room temperature, broad substrate scope. Readily available starting materials, such as alkenes and air, to construct valuable β-azido alcohols.

Wang *et al.* in 2020, demonstrated a metal-free method for the synthesis of β-trifluoromethyl hydrazines (**56**) by reacting alkene (**55**) with sodium trifluoromethanesulfonate as the CF_3 source. (**Figure 32**) [36]. This methodology enabled a radical cascade that incorporates a trifluoromethyl and a hydrazine group across the C=C double bond.

Moon *et al.* in 2020 demonstrated an atom-economical visible-light-mediated synthesis of aminopyridylationproduct (**56**) in the presence of alkenes (**55**) and *N*-aminopyridinium ylides (**Figure 33**) [37]. This environmentally friendly method

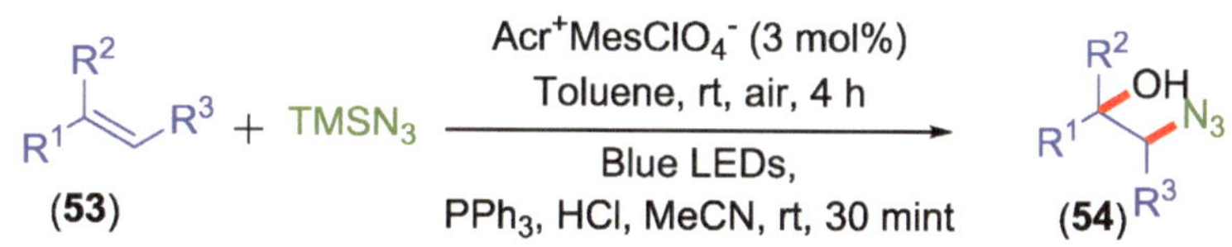

Figure 31.
Synthesis of Hydroxyazidation derivative.

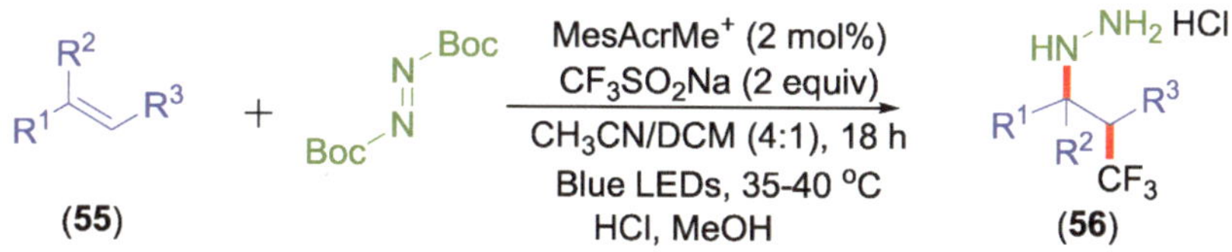

Figure 32.
Synthesis of β-Trifluoromethyl Hydrazines.

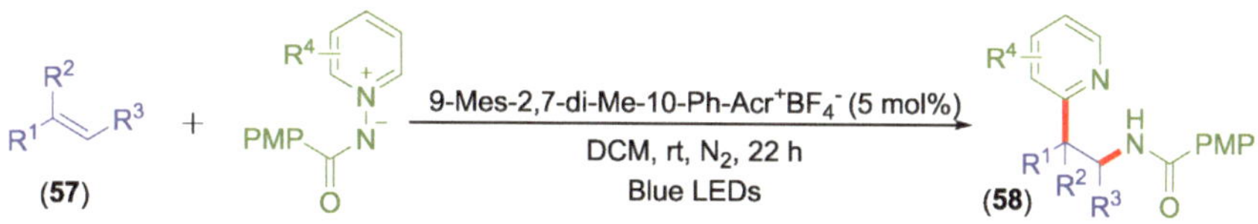

Figure 33.
Synthesis of Aminopyridylation product.

applies to a wide range of substrates with good functional group tolerance. Both activated, unactivated alkenes and pyridine are smoothly reacted and gave their desired products in moderate to good yields at room temperature.

5.3 Eosin-Y-catalyzed C-N bond formation

Moon *et al.* in 2019 demonstrated an Eosin Y mediated photocatalytic strategy for the synthesis of aminoethyl pyridine derivatives (**60**) in the presence of alkenes (**59**) using a variety of *N*-aminopyridinium salts as both aminating and pyridylating agents (**Figure 34**) [38]. Here concomitant incorporations of amino and pyridyl groups take place into alkenes under mild reaction conditions. In this protocol alkene bearing both electron-withdrawing and electron-donating groups are well tolerated. According to the possible mechanism (**Figure 34**) initially, in the presence of LEDs photocatalyst EY excited to the EY*, which subsequently form the unstable *N*-pyridine radical (A) via a single electron transfer (SET). Then the homolytic cleavage of the N–N bond takes place and generate the activated *N*-centered radical intermediate (B) and release the pyridine. The N-centered radical intermediate (B), attack the alkene and generate the intermediate (C). After that radical additions take place and produce the intermediate (D). Finally, the product (**60**) is obtained via deprotonations and the radical extrusion process.

Alam *et al.* in 2020 developed an elegant visible-light-mediated synthesis of *N*-hydroxybenzimidoyl cyanides from aromatic terminal alkenes using Eosin Y as a metal-free photocatalyst (**Figure 35**) [39]. DFT calculation supports a biradical pathway with successive incorporation of two nitrogen atoms, one each from *tert*-butyl nitrite (TBN) and ammonium acetate. The difunctionalization product is accomplished by the concomitant installation of an oxime and a nitrile group.

As determined from the DFT calculation and few control experiments a plausible mechanism has been proposed (**Figure 35**). In the influence of visible light Eosin Y (EY) undergoes excitation and generates a PINO radical from NHPI via hydrogen atom transfer (HAT) and returns to the ground state. The PINO radical adds to the

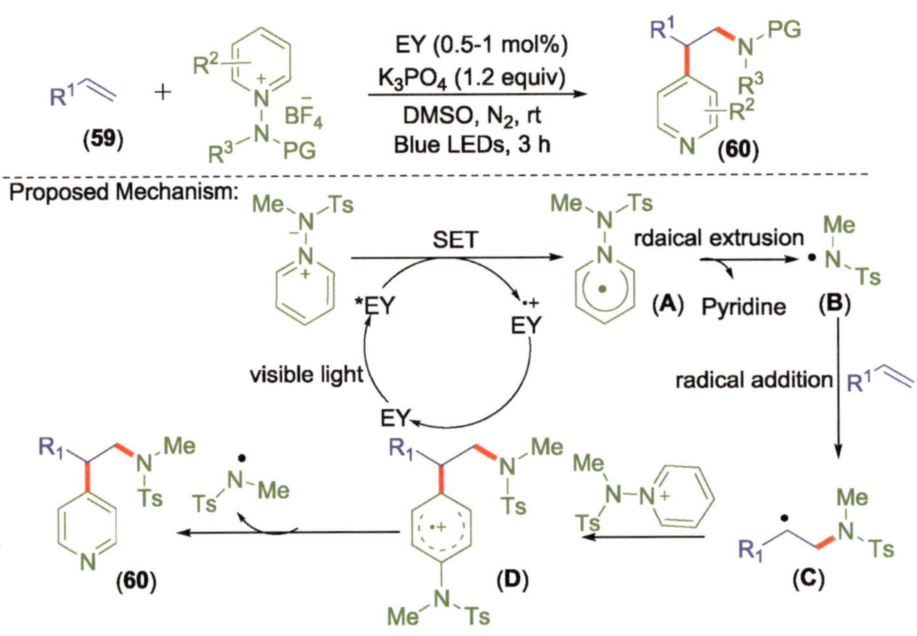

Figure 34.
Synthesis of Aminoethyl pyridine derivatives.

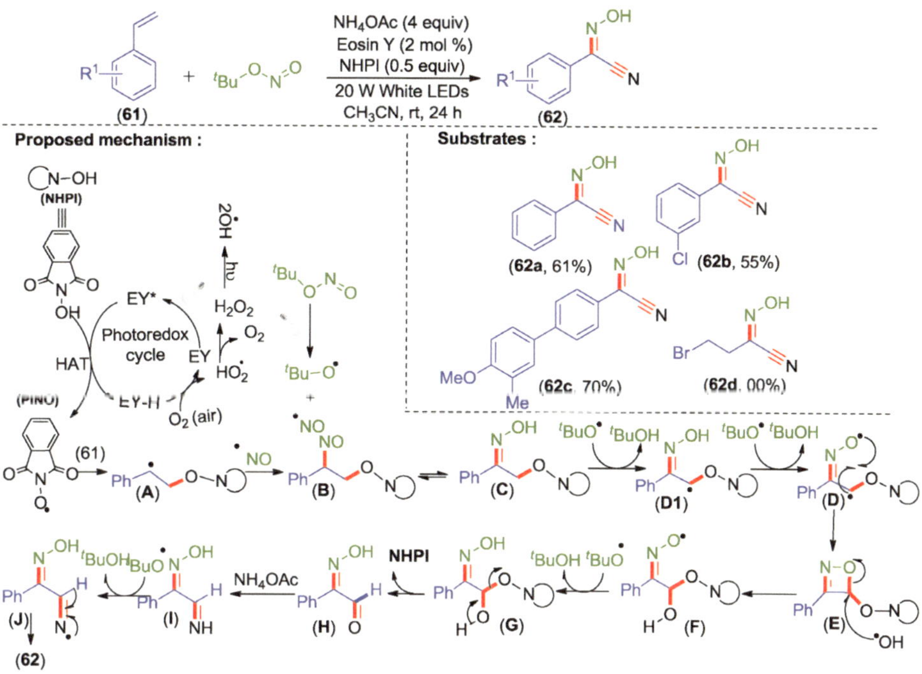

Figure 35.
Synthesis of N-*Hydroxybenzimidoyl cyanides.*

alkene (**61**) to give a benzylic radical intermediate (A) which trap the NO radical originating from TBN to give a nitroso intermediate (B) and is tautomerized to an oxime intermediate (C). Subsequent abstraction of two H atoms from the oxime intermediate (C) by in situ *tert*-butoxyl radical give a 1,4-biradical intermediate (D) which upon intramolecular coupling generates a four-membered cyclic intermediate (E). The strained cyclic intermediate (E) undergoes ring-opening via attack of an OH radical to form a hemiacetal radical intermediate (F). The N – O radical intermediate (F) abstracts a proton from *tert*-butanol to generate a neutral hemiacetal intermediate (G). The neutral hemiacetal intermediate loses NHPI providing an oxime aldehyde (H). Condensation between ammonia (generated from ammonium acetate) and the aldehydic intermediate (H) form an iminium intermediate (I). Abstraction of an iminium N – H from the intermediate (I) by the ^{t}BuO radical produce a nitrogen-centered radical (J). Finally, the abstraction of the aldehydic proton from intermediate (J) by *tert*-butoxy radical provided the cyano functionalized product (**62**).

6. Conclusion

In summary, this chapter focus on the recent advancements in visible-light-mediated transition-metal and organic dye catalyzed difunctionalization of alkene leading to the formation of C-N bond. The utilization of visible light by photo catalysis is a burgeoning field in contemporary organic synthesis The ubiquitous nature of the C-N bond predominates the synthetic chemist community. In this regard visible-light-mediated difunctionalization of alkene reactions have emerged as an efficient strategy for the synthesis of functionalized molecules, giving a high atom economy. Organic dye mediated C-N bond formations is even more promising compared to metal-catalyzed C-N bond formation because they overcome the drawbacks associated with transition metals that limits their use in pharmaceutical industries. With the current momentum of development, a greater impact of photocatalytic C–N bond-formations reactions is foreseeable, for example, in the late-stage modifications of natural products, large-scale syntheses and enantioselective C–N bond formations reactions Visible-light-mediated reactions provide a greener and sustainable approach and mild reaction condition towards the construction of complex molecules. Further developments in this area may open up broad opportunities for straightforward, efficient, and atom economical synthesis of N-compounds from simple alkenes.

Author details

Bhisma K. Patel*, Tipu Alam and Amitava Rakshit
Department of Chemistry, Indian Institute of Technology Guwahati, Guwahati, India

*Address all correspondence to: patel@iitg.ac.in

References

[1] Anastas PT, Warner JC, Green Chemistry: Theory and Practice, Oxford University Press, New York, 1998

[2] Nicewicz DA, MacMillan DWC. Merging Photo Redox Catalysis with Organocatalysis: The Direct Asymmetric Alkylation of Aldehydes. Science 2008; 322:77-80. DOI: 10.1126/science.1161976

[3] Yu Xiao-Ye, Chen Jia-Rong, Xiao Wen-Jing. Visible Light-Driven Radical-Mediated C-C Bond Cleavage/ Functionalization in Organic Synthesis. Chemical Reviews 2021; 121:506-561. DOI: 10.1021/acs.chemrev.0c00030

[4] Koike T, Akita M. A versatile strategy for difunctionalization of carbon-carbon multiple bonds by photo redox catalysis. Organic Chemistry Frontiers. 2016;3:1345-1349. DOI: 10.1039/C6QO00139D

[5] Chan Chun-Ming, Chow Yip-Chi, Yu Wing-Yiu. Recent Advances in Photocatalytic C–N Bond Coupling Reactions. Synthesis 2020; 52:2899-2921. DOI: 10.1055/s-0040-1707136

[6] Cao Min-Yi, Ren X, Lu Z. Olefin difunctionalizations via visible light photo catalysis. Tetrahedron Letters. 2015;56:3732-3742. DOI: 10.1016/j. tetlet.2015.04.091

[7] Koike T, Akita M. Visible-light radical reaction designed by Ru- and Ir-based photoredox catalysis. Inorganic Chemistry Frontiers. 2014; 1:562-576. DOI: 10.1039/c4qi00053f

[8] Lemos A, Lemaire C, Luxena A. Progress in Difluoroalkylation of Organic Substrates by Visible Light Photoredox Catalysis. Advanced Synthesis & Catalysis. 2019;361:1500-1537. DOI: 10.1002/adsc.201801121

[9] Dagousset G, Carboni A, Magnier E, Masson G. Photoredox-Induced Three-Component Azido- and Aminotrifluoromethylation of Alkenes. Organic Letters. 2014;16:4340–4343. DOI: 10.1021/ol5021477

[10] Yasu Y, Koike T, Akita M. Intermolecular Aminotrifluoromethylation of Alkenes by Visible-Light-Driven Photoredox Catalysis. Organic Letters. 2013;15: 2136–2139. DOI: 10.1021/ol4006272

[11] Carboni A, Dagousset G, Magnier E, Masson G. Photoredox-Induced Three-Component Oxy-, Amino-, and Carbotrifluoromethylation of Enecarbamates. Organic Letters. 2014; 16:1240–1243. DOI: 10.1021/ol500374e

[12] Yang Y, Xu Chong-Hui, Xiong Zhi-Qiang, Li Jin-Heng. Visible light photoredox alkylazidation of alkenes with sodium azide and heteroarenium salts: entry to azido-containing 1, 4-dihydropyridines. Chemical Communications. 2020;56:9549-9552. DOI: 10.1039/D0CC03235B

[13] Yu Xing-Long, Chen Jia-Rong, Chen Dong-Zhen, Xiao Wen-Jing. Visible-light-induced photocatalytic azotrifluoromethylation of alkenes with aryldiazonium salts and sodium triflinate. Chemical Communications. 2016;52: 8275-8278. DOI: 10.1039/C6CC03335K

[14] Govaerts S, Angelini L, Hampton C, Malet-Sanz L, Ruffoni A, Leonori D. Photoinduced Olefin Diamination with Alkylamines. Angewandte Chemie, International Edition. 2020;59: 15021-15028. DOI: 10.1002/ anie.202005652

[15] Hari DP, Hering T, König B. The Photoredox-Catalyzed Meerwein Addition Reaction: Intermolecular Amino-Arylation of Alkenes. Angewandte Chemie, International Edition. 2014;53:725-728. DOI: 10.1002/ anie.201307051

[16] Prieto A, Diter P, Toffano M, Hannedouche J, Magnier E. Photoredox-Initiated 1,2-Difunctionalization of Alkenes with *N*-Chloro *S*-Fluoroalkyl Sulfoximines. Advanced Synthesis & Catalysis. 2019;361:436-440. DOI: 10.1002/adsc.201801207.

[17] Ouyang Xuan-Hui, Li Y, Song Ren-Jie, Li Jin-Heng. Alkylamination of Styrenes with Alkyl *N*-Hydroxyphthalimide Esters and Amines by $B(C_6H_5)_3$-Facilitated Photoredox Catalysis. Organic Letters. 2018;20: 6659-6662. DOI: 10.1021/acs.orglett.8b02670

[18] Miyazawa K, Koike T, Akita M. Regiospecific Intermolecular Aminohydroxylation of Olefins by Photoredox Catalysis. Chemistry A European Journal. 2015;21:11677-11680. DOI:10.1002/chem.201501590

[19] Xu R, Cai C. Three-component difluoroalkylamination of alkenes mediated by photoredox and iron cooperative catalysis. Organic & Biomolecular Chemistry. 2019;17: 8541-8545. DOI: 10.1039/C9OB01815H

[20] Wu Y, Zhang Y, Jiang M, Dong X, Jalani HB, Li G, Lu H. Synergistic combination of visible-light photo-catalytic electron and energy transfer facilitating multicomponent synthesis of β-functionalized α,α-diarylethylamines. Chemical Communications. 2019;55: 6405-6408. DOI: 10.1039/C9CC02465D

[21] Chen J, Zhu S, Qin J, Chu L. Intermolecular, redox-neutral azidoarylation of alkenes via photoredox catalysis. Chemical Communications. 2019;55:2336-2339. DOI: 10.1039/C9CC00241C

[22] Guo W, Wang Q, Zhu J. Selective 1,2-Aminoisothiocyanation of 1,3-Dienes Under Visible- Light Photoredox Catalysis. Angewandte Chemie, International Edition. 2021;60: 4085-4089. DOI: 10.1002/anie.202014518

[23] Zong Y, Lang Y, Yang M, Li X, Fan X, Wu J. Synthesis of *β*-Sulfonyl Amides through a Multicomponent Reaction with the Insertion of Sulfur Dioxide under Visible Light Irradiation. Organic Letters. 2019;21:1935-1938. DOI: 10.1021/acs.orglett.9b00620

[24] Ge H, Wu B, Liu Y, Wang H, Shen Q. Synergistic Lewis Acid and Photoredox-Catalyzed Trifluoromethylative Difunctionalization of Alkenes with Selenium Ylide-Based Trifluoromethylating Reagent. ACS Catalysis. 2020;10:12414-12424. DOI: 10.1021/acscatal.0c03776

[25] Qin Q, Han Yue-Yue, Jiao Yan-Yan, He Y, Yu S. Photoredox-Catalyzed Diamidation and Oxidative Amidation of Alkenes: Solvent-Enabled Synthesis of 1,2-Diamides and α-Amino Ketones. Organic Letters. 2017;19:2909-2912. DOI: 10.1021/acs.orglett.7b01145

[26] Qin Q, Ren D, Yu S. Visible-light-promoted chloramination of olefins with *N*-chlorosulfonamide as both nitrogen and chlorine sources. Organic & Biomolecular Chemistry. 2015;13: 10295-10298. DOI: 10.1039/C5Ob 01725D

[27] Wu D, Cui Shuang-Shuang, Lin Y, Li L, Yu W. Visible Light-Driven Azidation/Difunctionalization of Vinyl Arenes with Azidobenziodoxole under Copper Catalysis. The Journal of Organic Chemistry. 2019;84: 10978-10989. DOI: 10.1021/acs.joc.9b01569

[28] Fumagalli G, Rabet PTG, Boyd S, Greaney MF. Three-Component Azidation of Styrene-Type Double Bonds: Light- Switchable Behavior of a Copper Photoredox Catalyst. Angewandte Chemie, International Edition. 2015;54:11481-11484. DOI: 10.1002/anie.201502980

[29] Hossain A, Vidyasagar A, Eichinger C, Lankes C, Phan J,

Rehbein J, Reiser O. Visible-Light-Accelerated Copper(II)-Catalyzed Regio- and Chemoselective Oxo-Azidation of Vinyl Arenes. Angewandte Chemie, International Edition. 2018;57: 8288-8292. DOI: 10.1002/anie.201801678

[30] Xiong Y, Ma X, Zhang G. Copper-Catalyzed Intermolecular Carboamination of Alkenes Induced by Visible Light. Organic Letters. 2019;21: 1699-1703. DOI: 10.1021/acs.orglett.9b00252

[31] Cheung KPS, Kurandina D, Yata T, Gevorgyan V. Photoinduced Palladium-Catalyzed Carbofunctionalization of Conjugated Dienes Proceeding via Radical-Polar Crossover Scenario: 1,2-Aminoalkylation and Beyond. Journal of the American Chemical Society. 2020; 142:9932-9937. DOI: 10.1021/jacs.0c03993

[32] Wei W, Cui H, Yue H, Yang D. Visible-light-enabled oxyazidation of alkenes leading to α-azidoketones in air. Green Chemistry. 2018;20:3197-3202. DOI: 10.1039/C8GC01245H

[33] Wang P, Luo Y, Zhu S, Lu D, Gong Y. Catalytic Azido-Hydrazination of Alkenes Enabled by Visible Light: Mechanistic Studies and Synthetic Applications. Advanced Synthesis & Catalysis. 2019;361:5565-5575. DOI: 10.1002/adsc.201901041

[34] Shen J, Xu J, He L, Ouyang Y, Huang L, Li W, Zhu Q, Zhang P. Photoinduced Rapid Multicomponent Cascade Reaction of Aryldiazonium Salts with Unactivated Alkenes and $TMSN_3$. Organic Letters. 2021;23: 1204-1208. DOI: 10.1021/acs.orglett.0c04148

[35] Yang B, Lu Z. Visible-Light-Promoted Metal-Free Aerobic Hydroxyazidation of Alkenes. ACS Catalysis. 2017;7:8362-8365. DOI: 10.1021/acscatal.7b02892

[36] Wang P, Zhu S, Lu D, Gong Y. Intermolecular Trifluoromethyl-Hydrazination of Alkenes Enabled by Organic Photoredox Catalysis. Organic Letters. 2020;22:1924-1928. DOI: 10.1021/acs.orglett.0c00287

[37] Moon Y, Lee W, Hong S. Visible-Light-Enabled Ortho-Selective Aminopyridylation of Alkenes with *N*-Aminopyridinium Ylides. Journal of the American Chemical Society. 2020;142: 12420-12429. DOI: 10.1021/jacs.0c05025

[38] Moon Y, Park B, Kim I, Kang G, Shin S, Kang D, Baik Mu-Hyun, Hong S. Visible light induced alkene aminopyridylation using N-aminopyridinium salts as bifunctional reagents. Nature Communications. 2019;10:4117. DOI: 10.1038/s41467-019-12216-3

[39] Alam T, Rakshit A, Begum P, Dahiya A, Patel BK. Visible-Light-Induced Difunctionalization of Styrenes: Synthesis of *N*-Hydroxybenzimidoyl Cyanides. Organic Letters. 2020;22: 3728-3733. DOI: 10.1021/acs.orglett.0c01235

Section 3

Olefins Synthesis

Chapter 4

Catalytic Isomerization of Olefins and Their Derivatives: A Brief Overview

Wangjing Ma, Bonan Liu, Duanda Wang, Jun Zhao, Lu Zhang and Lei Zhang

Abstract

Carbon–carbon double bond (CCDB) isomerization is a method for synthesizing new organic compounds from olefins and their derivatives, which was based on C=C migration along carbon chain and cis/trans transform, and it plays a vital role in the fields of organic synthesis, synthesis of daily chemicals, raw oil's development and synthesis of natural products and so on. In this paper, advances of five types of catalytic methods for CCDB of olefins and their derivatives since the 1960s were discussed in detail; Based on his recent work, the author mainly introduces the application and development of photocatalysis in CCDB of olefins and their derivatives.

Keywords: olefins, carbon–carbon double bond, isomerization, photocatalyst

1. Introduction

Olefins, also known as alkenes, are examples of unsaturated hydrocarbons and are made up of hydrogen and carbon atoms only and contain one or more pairs of carbon–carbon double bonds (CCDB). One CCDB consists of one sigma bond and one pi bond and is stronger than carbon–carbon single bond (611 $kJ \cdot mol^{-1}$ for C=C vs. 347 $kJ \cdot mol^{-1}$ for C-C), shorter than carbon–carbon single bond (average bond length: 0.134 nm for C=C vs. 0.154 nm for C-C). Depending on the position of the CCDB, olefins can be divided into terminal olefins, that the double bond is located at the terminal of a linear carbon chain (also called α-alkenes), and internal olefins, that the double bond is taking place at the inner side of carbon chain. It was valuable to study the effects of a CCDB's migration along the carbon chain on the hydrocarbons' activity and applications of a CCDB's migration along the carbon chain in oil drilling, surfactants, lubricants, fine chemicals, agrochemicals, pharmaceuticals, and other fields.

The following discussion will focus mainly on the isomerization of olefin, or migration of CCDB on linear alkenes. The functionalities of olefins, mainly for industrial and pharmaceutical purposes, vary by the location of CCDB on the alkene chains herein. Migration of CCDB is an important atom-economic reaction. New high valuable alkenes can be obtained from some inexpensive ones using the CCDB's migration along the carbon chain or cis/trans isomerism, as shown in **Figure 1**.

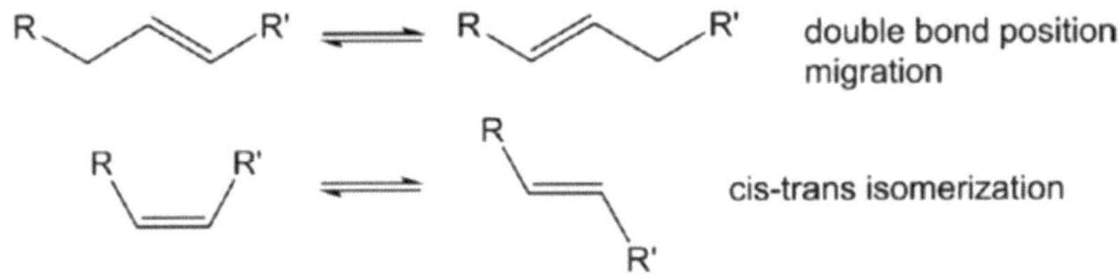

Figure 1.
C=C isomerization of alkenes. R, or R'= -H, -OH, alkyl, alkenyl, phenyl, cycloalkyl, cycloalkenyl, -CHO, -COOH, -COOR etc.

It is difficult for alkenes to synthesize positional or geometrical isomer by the conventional method, while relatively easier by isomerization [1]. In general, there are five methods to realize the olefine isomerization: acid catalysis, basic catalysis, molecular sieves, organometallic chain-walking catalysis, photocatalysis. These catalysis function mechanism varies, has the characteristic respectively. In the rest of the chapters, these five major categories of olefin isomerization methods will be explained in detail.

2. Acidic catalysis

The terminal linear alkenes generally range of C4 through C20 were isomerized to internal linear alkenes over a selective, strong acid catalyst to improve the positive CCDB isomerization and not catalyze oligomerization, cracking, skeletal isomerization, or alkylation. Firstly, the interaction of CCDB and acidic catalyst creates dissociation proton and then active carbonium ions, then induces dissociation among ortho protons; eventually generating new ectopic CCDB mixed olefin with certain cis/trans (Z/E) ratio in terms of thermodynamically stable.

Since the 1930s, Ipatieff et al. [2] studied isomerization of butene using various catalysts such as phosphoric acid, perchloric acid, toluene-p-sulfonic acid and zinc chloride solutions, shown in **Figure 2**. In the 1960s, F. Asinger and his co-workers [3] catalyzed isomerization of α-undecene to form internal undecene with homogeneous catalyst like sulfuric acid, perchloric acid and other moderately strong acid, respectively, shown in **Figure 3**.

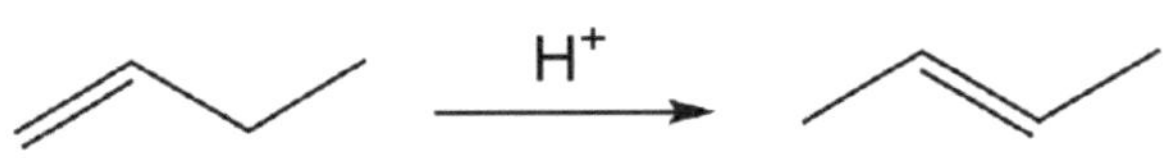

Figure 2.
C=C isomerization in butene.

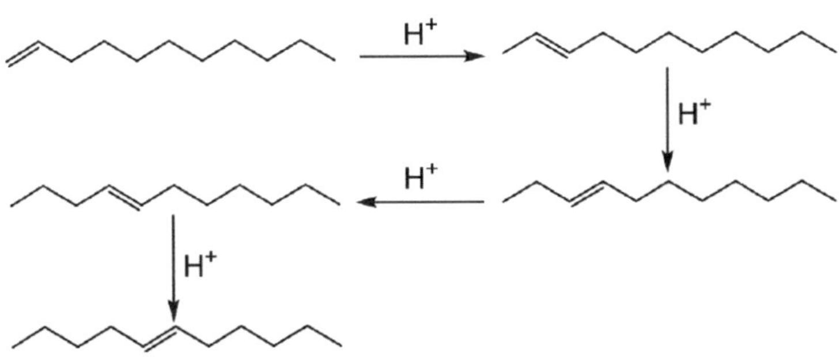

Figure 3.
C=C isomerization in α-undecene.

α-alkenes can be isomerized into mixed internal olefins isomers under appropriate solvents systems, acidity, temperature, and other reaction conditions that enhance catalysis. As a catalytic system, perchloric acid/anhydride can promote isomerization of α-undecene under 100°C in a high conversion rate with few side reactions (mainly oligomerization) [3]. Other Lewis acids such as boron (tri)fluoride (BF_3), aluminum chloride ($AlCl_3$), and borofluoric acid (HBF_4) were used to catalyze α-undecene to lead to more α-undecene oligomerization in even less time. However, acidic catalysts used inhomogeneous reactions like those mentioned above are hard to recycle and high in attrition rate. Therefore, recent research mainly focuses on using solid acids Al_2O_3, SiO_2, WO_3, Al_2O_3-SiO_2, mesoporous Ti, Nb and Ta oxides, etc., on performing the acidic olefin isomerizations, which are heterogeneous catalysts [4–8]. Their catalysis efficiencies depend on their acidity in general. To increase the surface acidity and catalytic activities of these heterogeneous catalysts, researchers covered them, aluminum oxides/salts or silicon/silica, by a various set of materials as follows: (1) proton acids (sulfuric acid, phosphoric acid, hydrochloric acid, nitric acid, boric acid, hydrofluoric acid); (2) Lewis type acids ($AlCl_3$, SbF_5, BF_3, and P_2O_5); (3) ammonium salts (NH_4F). (4) Strong chlorinating agents (PCl_3 and PCl_5); (5) a strong Lewis acid like CH_3AlCl_2; (6) molecular halogens (Cl_2 and I_2) [4].

$AlCl_3$, as a Lewis acid, is a problematic catalyst for CCDB isomerization of alkenes which forms more by-product, requires large dosage of catalysts and hard to recycle. Acid sites in γ-Al_2O_3 as an amphoteric oxide can be used as catalytic center to catalyze isomerization of α-molecule to promote CCDB's migration and cis-trans isomerism, while non-acidic sites in catalysts can promote other side reactions such as alcohol dehydration; As a heterogeneous catalyst, SiO_2-Al_2O_3-MgO promoted 1-butene isomerized to both cis- and trans-2-butene with the isomerization taking placing via π-complexes on the acidic sites of the catalysts, which are easy to be separated from reaction system, thus gradually replacing the homogeneous catalysts to be applied in olefin isomerization (**Table 1**).

Among metals oxide listed in **Table 2**, either sulfated mesoporous Nb oxides ($C_{12}H_2SO_4$ Meso Nb) or Ta oxides ($C_{12}H_2SO_4$ Meso Ta) showed higher activities and selectivity than sulfated mesoporous Ti oxides ($C_{12}H_2SO_4$ Meso Ti), Amberlyst 15, HY zeolite and H-ZSM5 in CCDB isomerization of 1-hexene. The conversion rate of 1-hexene to trans/cis 2-isomers reaches 95.89% when using $C_{12}H_2SO_4$ meso Ta as catalysts in 4 hrs was reported. And the ratio of trans/cis isomers reaches up to 3.7 after 6 hrs, shown in **Figure 4**. $C_{12}H_2SO_4$ meso Ta showed both high activity and selectivity, which can be attributed to its high BET surface area (292.19 $m^2 \cdot g^{-1}$),

Sample	pH	Acid amount ($mmol \cdot g^{-1}$)
C_{12}Meso Ti	+3.3	2.0
$C_{12}H_2SO_4$ Meso Ti	+0.8	4.7
C_{12}Meso Nb	−6.6	2.4
$C_{12}H_2SO_4$ Meso Nb	−8.2	31.7
C_{12}Meso Ta	−6.6	0.4
$C_{12}H_2SO_4$ Meso Ta	−8.2	19.8
HY Zeolite	−6.6	1.5
H-ZSM5	−4.4	16.1
Amberlyst 15	N/A	N/A

Table 1.
Acid strength and acid amount of solid acid catalysts (measured by Hammett indicators and n-butylamine titration) [9].

Material	Xi (%)	Si (%)	$TOF \times 10^3 (s^{-1})$
SAC-13	82	65	36
XZO 1251	51	55	27
XZO 1251[a]	98	−30	120
SAPO-11	74	83	2.3
ZSM-35	74	66	4.1
K2620	79	70	0.81
AM-15[b]	23	78	0.41
AM-35	74	76	0.75
AM-70	7.1	24	0.062
AM-XN1010	74	43	1.3
BCPR4	4.2	31	1.3
BCPR5	15	74	4.2
BCNA2	0	N/A	0

[a]185°C
[b]110°C
Xi: Conversion to internal alkene;
Si: Selectivity to internal alkene;
TOF: Turnover frequency based on the measured acid site densities are defined by Eqs. (1)–(3).

$$Xi = \left(1 - mol\left(\frac{\text{Terminal, Product}}{\text{mol(Terminal, Feed)}}\right)\right) \times 100\% \quad (1)$$

$$Si = \frac{\text{moles(Target, Product)} - \text{moles(Target, Feed)}}{\text{moles(Target + Branched + 2} \times \text{Dimer, Product)} - \text{moles(Target + Branched + 2} \times \text{Dimer, Feed)}} \times 100\% \quad (2)$$

$$TOF = \frac{Xi(\%)}{100\%} \times \left(\frac{mLC_{16}}{min}\right) \times \left(\frac{0.783gC_{16}}{mLC_{16}}\right) \times \left(\frac{molC_{16}}{224gC_{16}}\right) \times \left(\frac{g\ catalyst}{meq}\right) \times \left(\frac{1}{g\ catalyst}\right) \times \left(\frac{1000\ meq}{mol}\right) \times \left(\frac{1\ min}{60s}\right) \quad (3)$$

*The TOFs in **Table 2** were determined from the calculated first-order rate constants, although in practice these numbers are very similar to what is found using Eq. (3).*

Table 2.
Isomerization of α-hexadecenes catalyzed by solid acid catalysts listed [10].

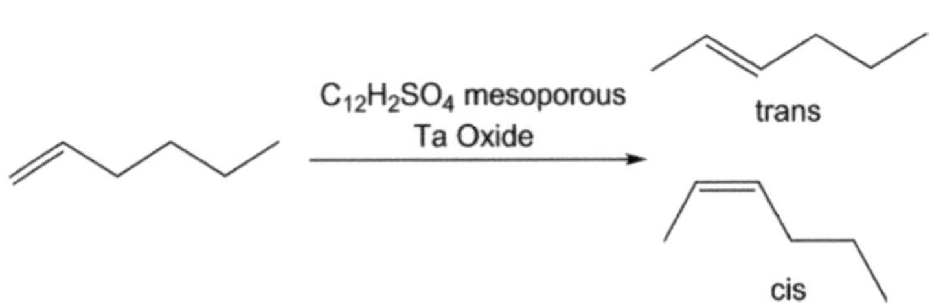

Figure 4.
C=C isomerization in α-hexene catalyzed by $C_{12}H_2SO_4$ mesoporous Ta oxide.

optimal pore size (18.2 Å), and increased concentration of active Brønsted acid sites on the surface of the mesoporous channels [8].

Heteropoly acids (HPAs) are a unique type of materials that are active both in oxidation–reduction and acid catalysis [9, 11, 12]. The polyoxometalates form heteropoly anions with metal-oxygen octahedra and work as the basic structural unit. The Keggin-type HPAs are the most important in catalysis: the Keggin heteropoly anion has the empirical formula, $XM_{12}O_{40}n^{-}$, wherein X is a non-metal heteroatom (e.g. P^{5+}, Si^{4+} or B^{3+}) and M is a metal addenda atom (e.g. MO, W, V, Ce, Zr, Nb, Sb and Ti). Exemplary Keggin heteropoly acids in which X is phosphorus. 12-tungstophosphoric ($H_3PW_{12}O_{40}$, TPA) is the most usual catalyst of choice because of its high acidic strength and relatively high thermal stability. They are strong Brønsted acid catalysts, and are stronger than conventional solid acids like zeolites and mixed oxides.

HPAs can be used either directly as a bulk material or with the supports. The supported form is preferable because of relatively higher surface area compared with the bulk material (5–8 $m^2 \cdot g^{-1}$) and better accessibility of reactants to the active sites, on the premise that carriers have little interaction with HPAs. Solid carriers, including acidic (silica, acidic ion-exchange resin) and neutral solids carriers (active carbon), have been reported that are suitable as HPA supports. The molecular structure is shown in **Figure 5**.

Zhiping Du et al. [12] used HPAs supported by Al_2O_3 to catalysis on isomerization of 1-hexene to 2-hexene, 3-hexene, and 1-octene to 2-octene, 3-octene, 4-octene, etc. They investigated the influence of catalyst dosage, time, temperature, reactant purity, activation conditions of the HPAs, and supporting materials on the isomerization efficiency of 1-hexene and 1-octene. Their study showed that HPAs could promote olefin isomerization under a low temperature that CCDB migration and cis-trans isomerization take the main place with little skeleton isomerization.

However, HPAs catalysts for CCDB isomerization of alkenes are not perfect and do have it own problems. A typical issue is the thermal stability of HPAs wasn't high enough for conventional regeneration by the burning of coke at 500–550°C as routinely used in the case of zeolites and aluminosilicates because susceptibility of these types of catalysts to deactivation during organic isomerization reactions due to the formation of carbonaceous deposits (coke) on the catalyst surface.

Cation exchange resin (CER) [13–20] is a type of solid catalysts in which its catalytic mechanism is the same as homogeneous catalysts, while it does have active sites like silica gel, Al_2O_3, and zeolite as carriers, brings to CER become a pseudo homogeneous system. CER is used to catalyze gaseous short-chain alkene isomerization ($C_4 \sim C_7$). For example, α-butene could be isomerized to form certain cis/trans ratio 2-butene with yields up to 90% [10, 21–29]. In general, advantages of

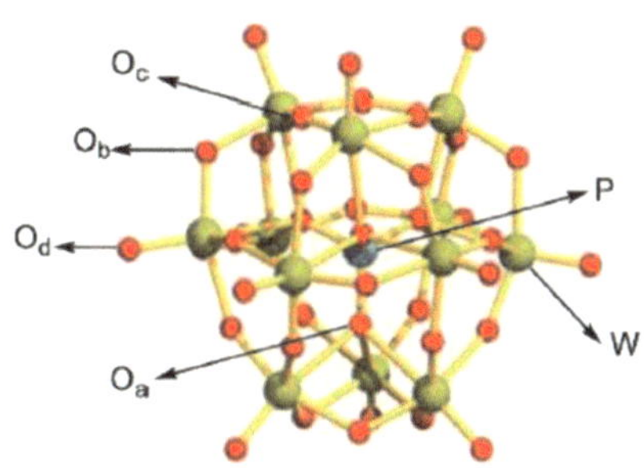

Figure 5.
The structure of 12-tungstophosphoric acid.

CER are: (1) it can be made in different shapes, structures and load capacities for different application purposes; (2) active groups are usually located on the surface of internal holes of CER, which is easy to contact with olefin molecular with faster reaction rate and higher yields; (3) catalysts are pseudo homogeneous, which means separation between catalysts, reactants and products are easier than homogeneous catalysts, for instant, homogenous acid catalysts have to deacidify after the reaction; so that bring to (4) catalyst may be reused for several times to avoid waste liquid's pollution to environment.

3. Basic catalysis

In base catalysis of olefin isomerization, α-alkenes or their derivatives (1) form secondary (or tertiary) carbanions with the base anions (B^-), then (2) rearrange into more stable primary carbanions, which (3) react with base (BH) to form new internal alkenes or their derivatives.

Forming cis-isomers is preferred in base-catalytic isomerization because the allyl intermediate is generated by losing protons from the olefin, and cis-isomer is more stable than trans-isomer.

The Shell Higher Olefin Process (SHOP) is a well-known and important base catalytic method that the Royal Dutch Shell commercialized in 1977 to produce α-olefins with controllable length from ethylene and subsequently to manufacture the corresponding aldehydes and fatty alcohols for producing detergents and surfactants [30]. The SHOP process incorporates the following reactions: oligomerization (**Figure 6**) [30], double bond isomerization (**Figure 7a**), and metathesis (**Figure 7b**) [30–32].

Some long chain alkenes isomerization, like isomerizing α-$C_{34}H_{68}$ to 11-$C_{34}H_{68}$, is industrially accomplished by Na/K loaded Al_2O_3 catalyst, or MgO, under reaction temperature at 80–140°C and pressure at 0.34–1.72 MPa [30–32]. The conversion rate of α-$C_{34}H_{68}$ into 11-$C_{34}H_{68}$ is more than 90%. Then subjected to the metathesis reaction, and in which catalysts comprising an alkali metal such as sodium, potassium dispersed on a high surface area, mainly inert, solid supporter, such as Al_2O_3 were prepared at the temperature between 5 and 50°C and in the molecular-oxygen-containing activating gas of oxygen-to-alkali ratio of 0.01–2 [31]. In the other case, catalyst on CCDB isomerization were those that have little

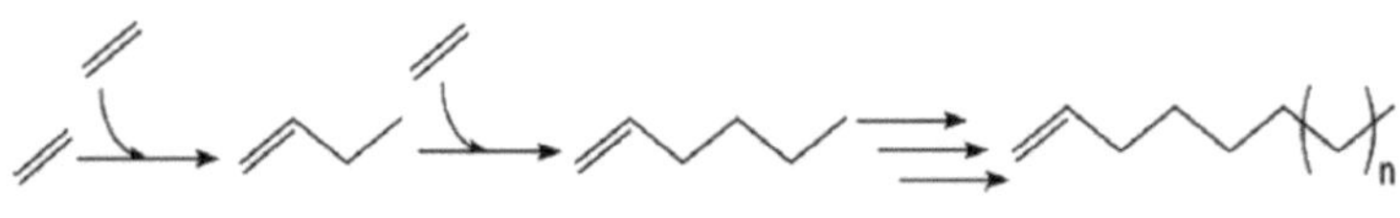

Figure 6.
Oligomerization of ethylene by Ni catalysts under 80–120°C, and 1000-2000psig.

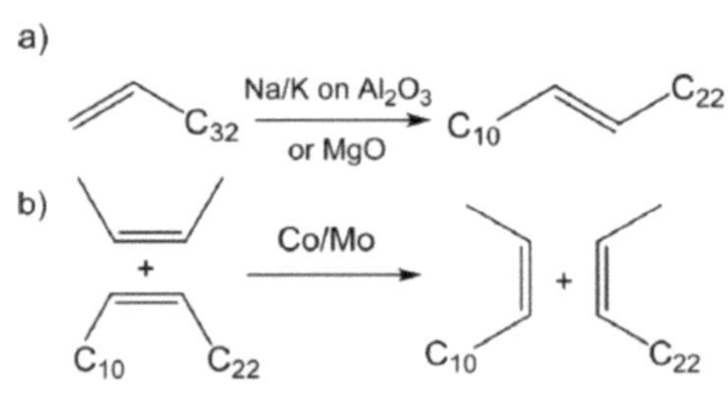

Figure 7.
SHOP process steps: a) alkene isomerization and b) metathesis.

polymerization and cracking, and include supported bauxite, alumina supported cobalt oxide or iron oxide or manganese oxide, and so on [32].

Under the conditions like that of isomerization, higher molecular like 11-$C_{34}H_{68}$ reacted with lower molecular like butadiene in metathesis reaction to produce a mixture of olefins such as internal $C_{13}H_{26}$, and $C_{25}H_{50}$, which are for chemical intermediates required by surfactants and detergents. The oligomerization of ethylene by Ni catalysts in the SHOP process was showed in the follows as **Figure 8** [18].

The industrial manufacturing of co-monomer ethylidene norbornene (ENB) also utilizes the alkene isomerization [33]. The reaction pathway is using Na-loaded-Al_2O_3 catalyst to the vinyl norbornene, produced from a Diels-Alder reaction, which deprotonated, rearranged, and reprotonated into the ENB (**Figure 9**).

Sumeet K. Sharma et al. [34] studied the selective double bond isomerization of allyl phenyl ethers catalyzed by ruthenium metal complexes, achieve a conversion rates of methyl chavicol (99.7%) with 95.4% selectivity of trans-anethole and eugenol (99.8%) with 95.6% selectivity of trans-isoeugenol in ethanol using $RuCl_2(PPh_3)_3$ catalyst. Ruthenium catalysts [35, 36] are relatively expensive, although they present an extraordinary performance on catalysis efficiency, the potential of massive production are very gloomy. On the other hand, inexpensive

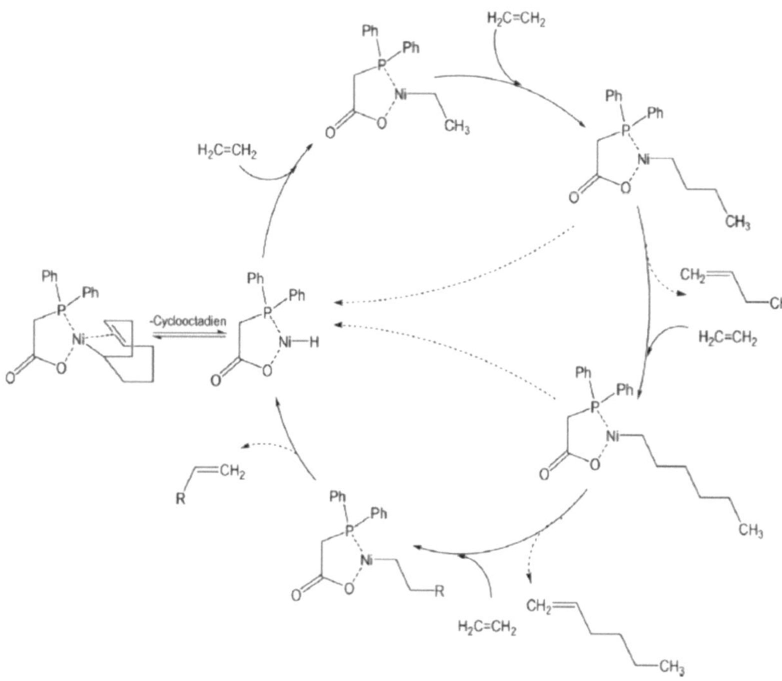

Figure 8.
The oligomerization of ethylene by Ni catalysts of the SHOP process.

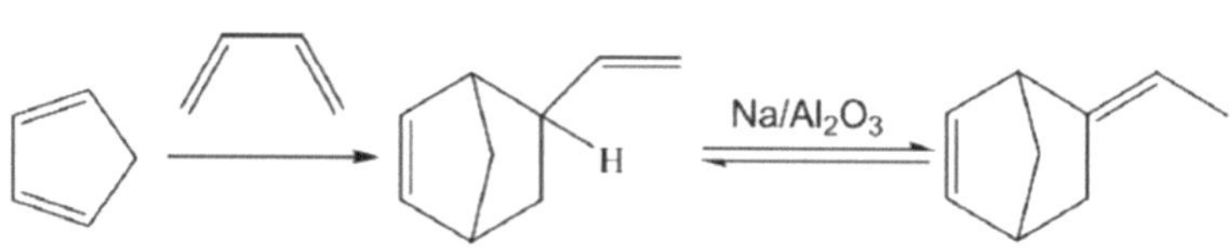

Figure 9.
Ethylidene norbornene formation by catalyzed isomerization of vinyl norbornene.

transition metals catalysts usually have a poorer performance, but it is more likely for industrial scale production after enhancing their efficiency.

Jiraporn Puriwat et al. [37] studied the gas-phase isomerization of 1-butene to 2-butene on the $MgO/Mg(OH)_2$ catalysts containing different basicity sites (i.e. weak, medium, and strong basic sites). The isomerization reactions mainly occur on either the strong or the medium-strength basic sites. In fact, the weak basic sites had little impact on the isomerization activity. The crystalline MgO contained both weak and strong basic sites, while the Mg $(OH)_2$ phase exhibited only the medium strength sites. In the lattice of crystalline MgO, strong basic sites are coming from the presence of oxygen atoms. The medium strength basic sites in Mg $(OH)_2$ risen from the hydroxyl groups in the Mg $(OH)_2$ structure.

Some strong base, like KOH [38], can be used independently in DMSO or alkoxide system, to catalyze position isomerism of propenyl C=C, to produce the intermediate could be applied in the perfume, cosmetic, pharmaceutical, and materials chemistry, and also as intermediates in synthetic sequences for the construction of more complex products. KOH, or NaOH was also a catalyst on C=C isomerization of allyl aromatics estragole and eugenol. Allyl aromatics estragole and eugenol could further turn into the corresponding alkenyl aromatics trans-anethole and trans-isoeugenol, respectively. Under certain conditions, 56% conversion was achieved with modest E:Z in 12 hours. This methodology does have drawbacks including the following as: the use of strong base in stoichiometric amounts, longer reaction times, incomplete conversion of reactant, low selectivity for the trans-isomer, high temperatures, caustic waste, and the likely need to separate the cis-isomer [39, 40].

4. Molecular sieves

According to the carrying amount of acid sites or alkaline sites, the molecular sieve can be divided into acid catalyst and base catalyst. Due to their unique structures and a broad prospect in industrial application, taking molecular sieves as an option is unavoidable for studies on catalysis that would like to apply their founds on an industrial scale. For the CCDB isomerization of alkenes, molecular sieves can be further subdivided into zeolite and non-zeolite catalysts. Non-zeolite molecular sieve are those formed by metal skeleton compound MOF, SAPO, etc. [41, 42].

Hu etc. al. [40] reported that mesoporous molecular sieve supported vanadium catalysts such as V-MCM-41, V-SBA-15 and V-TUD-1 with different pore structures and pore sizes were prepared via a post-synthesis grafting method using atomic layer deposition, by means of which the isomerization of 1-heptene was employed as a reaction probe to characterize the acid properties of these catalysts with moderate acid strengths. The results were that conversion yield as high as 92% at 600 K and isomer selectivity over 90% (mainly double bond shift products) over the V-SBA-15 catalyst were achieved.

Gajda et al. [41–45] developed a series of non-zeolite molecular sieve (NZMS) in succession in 1990s. The composition of synthetic molecular sieves, pore structure, and active component distribution at molecular level is controllable. It successfully catalysis CCDB isomerism of α-butene, α-pentene and other short-chain olefin, to produce tertiary olefins, which can be used to make high octane ether compounds required by reformed gasoline. The preferred NZMS were the silicoalumino-phosphate molecular sieves described in the patent [42], which are disclosed as microporous crystalline silicoaluminophosphates. They have a three-dimensional microporous framework structure of PO_2^+, AlO_2^- and SiO_2 tetrahedral units, and whose empirical chemical composition on an anhydrous basis is:

$$mR:\ (Si_xAl_yP_z)O_2 \tag{4}$$

wherein "R" represents at least one organic templating agent presented in the intracrystalline pore system; "m" represents the moles of "R" present per mole of $(Si_xAl_yP_z)O_2$ and has a value of from 0.02 to 0.3; "x", "y" and "z" represent the mole fractions of silicon, aluminum, and phosphorus present in the oxide moiety. Gajda et al. [46, 47] further developed a method to increase the proportion of isobutene with less by-products, by using a catalyst comprising at least one NZMS, which contains framework tetrahedral units (TO_2) of aluminum (AlO_2), phosphorus (PO_2), and at least one additional element (EL) as a framework tetrahedral unit (ELO_2).

A NZMS made up by Ferroaluminophosphates are also presented in patent [48], which has a three-dimensional microporous crystal framework structure of AlO_2, FeO_2 and PO_2 tetrahedral units, and whose empirical chemical composition on an anhydrous basis is:

$$mR:\ (Fe_xAl_yP_z)O_2 \tag{5}$$

Or the NZMS were crystalline microporous aluminophosphates in which the substituent metal is one of a mixture of two or more divalent metals of the group magnesium (Mg), manganese (Mn), zinc (Zn) and cobalt (Co), presented in patent [49], whose empirical chemical composition on an anhydrous basis is:

$$mR:\ (M_xAl_yP_z)O_2 \tag{6}$$

Or the NZMS were the MnAPSO molecular sieves [50] which have a framework structure of MnO_2, AlO_2^-, PO_2^+ and SiO_2 tetrahedral units whose empirical chemical composition on an anhydrous basis is:

$$mR:\ (Mn_wAl_xP_ySi_z)O_2 \tag{7}$$

Zeolite catalysts are aluminosilicate salts, depending on their pore structures and adjustable acidities, they could be divided into X-, Y-, ZSM-5, β-type, etc. [51, 52].

5. Organometallic "chain-walking" catalysts of transition metal

Depending on the element they used, transition metal catalysts can be classified as noble metal catalysts or non-noble metal catalysts. Noble metals, by definition, are metallic elements distributed in group VB, VIB, VIIB, and VIII. Typical examples are Pd, Ru, Rh, Ir, Os, etc. [38, 53–60]. Their complex compounds catalyze the isomerization by converting the double bonds of the α-alkenes to the β and γ positions, which is different from the acidic catalyst that catalyzes the isomerization reaction by converting the double bonds of both α- and β-alkenes to internal positions further. It also promotes side reactions that include cracking, oligomerization, skeletal isomerization, and alkylation.

Noble metal catalysts, however, are expensive and unrecyclable, which limited their application in the industry. While some common metals like Fe, Cu, Zn, Co, etc. [38, 61–64] are much cheaper and can promote C=C migration of α-linear olefins to generate internal olefins in high efficiency, too, providing another option. For example, $CoCl_2$ and Grignard reagent can isomerize α-tetradecene to generate (E)-2-, (Z)-2-, 3-tetradecene, and other isomers under 50 Celcius degree [38]. Also, Fe and Co complex compounds can shift the position of the C=C on the allyl benzene side chain, isomerized, and generated into various allyl benzene intermediate. These can be used as intermediate of perfume, antibiotics, insecticides, anti-Leishmania drugs, antifungal, algae inhibitors, anti-inflammatory drugs, antioxidants etc., which leads to a strong industrialized potential.

6. Photocatalysis

In 1983, the photocatalytic carbonylation synthesis reaction of aromatic halogenated hydrocarbons was realized, leads the dawn of the application of photocatalysis in organic synthesis. In the few decades, photocatalytic ring-opening polymerization and photocatalytic epoxidation of olefins have been reported, which makes photocatalytic organic synthesis has become an important branch in the field of photocatalysis. Comparing to the thermal catalysis, photocatalysis reactions are relatively environmental friendly, requiring more mild reaction conditions, different selectivity of isomers' structures. These advantages provide some special merits that brings more options when chooses the reaction routes for both academic studies and industrial manufacturings.

Cirjak and his co-workers [65] used metal clusters $(C_5R_5)_aFe_bM_c(CO)_dL_e$ as the photocatalyst to catalysis isomerism of olefin, in which the catalyst's absorption wavelength covers up to 220 nm with a low yield. Except for catalysis on CCDB position migration isomerism of alkenes, some photocatalysts such as Pd@TiO_2, could shift the allylic C=C position of the allyl aromatics along allyl chain to form isomers [66]. $[HFeM(CO)_8L^-]$ or $Fe(CO)_5$ can be used as photocatalyst in THF to catalyze 1-allylbenzene at room temperature under illumination conditions, isomerized into 2-allylbenzene [67]. Under mild conditions, the reaction productivity is over 98%, and the E/Z ratio in product is 10:1 which is a decent selectivity.

Recent years, Ma et al. [68] discovered a series of solid heteropoly acids and their supported zeolite sieves as photocatalyst to catalysis on CCDB isomerization of alkenes. A small amount of catalysts shows a strong catalytic activity on C=C isomerization of linear chain α-olefin (C_5-C_{20}) into internal olefins under visible light, in which the conversion rate can be up to 80% in 1 hour, and presents a high isomer selectivity (either E or Z). It was perfectly avoiding the complicated separation of E&Z isomers which is important for some purposes, like pharmaceutically, which do require isomer selectivity. Under ultraviolet irradiation, the catalysts to oleic acid ($CH_3(CH_2)_7CHCH(CH_2)_7COOH$, cis), priority to generate 3-trans-, 6-trans-, 11-trans-octadecenic acid and so on, with the conversion rate being up to 80% in 1 hour. The catalytic system also can catalyze a series of phenylpropyl compounds to make CCDB isomerization under ultraviolet irradiation, mild temperature and ordinary pressure to produce certain location isomer, which were widely used in production of high value chemical products applied in the field of medicine, biology and materials science, as shown in **Figure 10**.

This type of multifunctional catalyst combined four catalysis: photocatalysis, acid catalysis, molecular sieve catalysis and transition metal catalyzers which synergistically worked to promote C=C double bond migration of olefins and its derivatives. The main problems are as following: narrow catalysis absorption wavelength range, which was mainly in ultraviolet and near ultraviolet regions, low utilization of solar energy; the recombination rate of photo-generated carriers and hole is high, which results in lower quantum efficiency. Thus, we need to modify the structure and component of catalysts at present, such as the control of catalyst's crystal structure and defects, adjusting the energy band location and surface photosensitization etc.

However, photocatalysis on olefin's isomerization is not well-known enough which has to clarify the photocatalytic reaction mechanism, especially photo

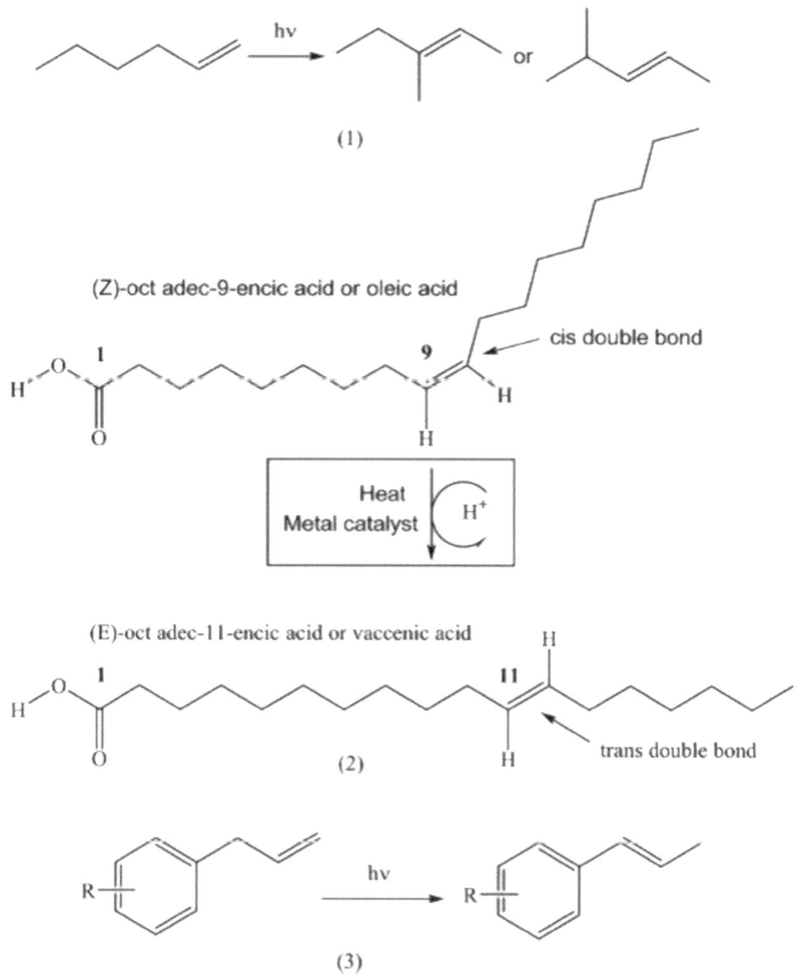

Figure 10.
CCDB isomerization of straight chain alkenes (1), oleic acid (2) and Allyl benzene (3).

generated carrier separation, transfer and interface shift process, that are essential to further improve the catalytic efficiencies. On another hand, future works may take advantages from some well-studied photocatalysis fields like nanomaterials, semiconductor physics, and photocatalysis organic reaction, to produce novel photocatalytic materials with high efficiency and high catalytic activity.

7. Conclusions

In this paper we summarized olefin isomerization catalysis in five categories, acidic, basic, molecular sieve, transitionmetal catalysis, and photocatalysis, and under each category, we present some typical catalysts that have been studied. The five categroies differentiates most available catalysis systems by their reaction mechanisms, and they also reflect somehow their properties, like acidic catalysis usually have issuses on products selectivity and catalysts recycling. The **Table 3** followed presents all catalysis systems mentioned before, what they are, which category they belong to, applicable reactants and desire products, with their purposes and features, pro and cons, and references of the works.

Method	Catalysis System	Olefin and Derivatives	System Features	References
Acidic Catalysis	H_2SO_4, HNO_3, HCl, H_3PO_4, p-H_3C-C_6H_4-SO_3H, C_6H_5-SO_3H	Linear α-olefins (C_4-C_{20})	Brønsted Acid, homogeneous catalysis, high reaction rate, conversion rate, selectivity; Poor to recycle, corrosive	[3]
	BF_3, HBF_4, Tetramethylene sulfone		Lewis Acid, high reaction rate and conversion rate; may causes polymerization of olefins; highly toxic, inflammable, poor to recycle, corrosive	[3]
	$AlCl_3$		Mild reaction rate, conversion rate, and selectivity, inexpensive; may causes polymerization of olefins; heavy dosage, poor to recycle	[3]
	Sulfided ion exchange resin		Gas-phase reaction, isomerizing both cis- and trans-β olefins; incapable of long liner olefin isomerization	[5–8, 11, 25–29]
	Cation exchange resin, perfluorinated ion exchange resin		Robustness to deactivate, longevity; slow reaction rate, side reactions	[10, 13–18]
	Al_2O_3, SiO_2, WO_3, ZrO_2, TiO_2, Nb_2O_5, Ta_2O_5, and heteropoly acids	Linear α-olefins (C_4-C_7)	Lewis Acid, slow reaction rate, conversion rate, poor selectivity, heterogeneous catalysis. Heteropoly acids	[21–24]
Basic Catalyst	Na/K-Al_2O_3, MgO	α-$C_{33}H_{66}$	The Shell Higher Olefin Process (SHOP) industrial catalyst, reaction isomerized 11-$C_{33}H_{66}$, conversion rate>90%	[30]
	γ-Al_2O_3-NaOH-Na, γ-Al_2O_3-Na_2CO_3-Na	Linear α-olefins	Good reaction rate, conversion rate, and selectivity; high temperature required	[30]
	Na-Al_2O_3, Potassium tert-butoxide, Ti-Zeigler (X_3TiH)	Vinyl Norbornene	Ethylene C=C positional transfer, industrial catalyst in use	[35]
	Strong base like KOH; KOH/DMSO; other hydroxides/alcohol-salt system	Propenyl aromatic compounds	Promote internal movement of C=C on propylene to form allyl compound; wild applied in perfume and food industry; long reaction time, low conversion rate, hard to purify	[38]
Molecula Sieve Catalysis	$0.83Na_2O$-$1.00Al_2O_3$–$2.48SiO_2$	α-olefins (C_6~C_{25})	>70% conversion rate in short time under room temperature; low side reaction; longevity catalyst lifetime	[39]
	Mesoporous molecular sieve like MCM-41, SBA-15, and TUD-1		92% conversion rate and 90% selectivity using V-SBA-15 to catalysis α-octene isomerization under 600 K	[40]
	Non-zeolite molecular sieve(MZMS); MOF-mR: $(Si_xAl_yP_z)O_2$	1-butene	Synthesis isobutene and 2-butene under 200~600°C	[41]

Method	Catalysis System	Olefin and Derivatives	System Features	References
	Non-zeolite molecular sieve(MZMS); MOF-mR: $(Si_xAl_yP_z)O_2$ MZMS mR: $(M_xAl_yP_z)O_2$, M=Co, Fe, Zn, Ca, Mn, etc.	1-butene 2-butene	Synthesis isobutene and 2-butene under 200~600°C Synthesis 1-butene and isobutene, could be further used to obtain ether for reformulated gasoline	[43]
	NZMS	1-butene 1-pentene	Isomerizing into isobutene, few side products.	[44–49]
	NZMS SAPO-11	1-hexene	Reinforce positional isomerism instead of cis-trans isomerism	[50]
	Zeolite molecular sieve: X, Y, ZSM-5, etc.		Isomerizing 1-hexene into 2- and 3-hexene	[51, 52]
Transition Metal Catalysis	Pd black/$PdCl_2$	Linear α-olefins, ketenes	A industrialized catalyst to isomerizing α-olefin into stable, internal form. Expensive, toxic, complicate for further processing	[53]
	$Ru_3(CO)_{12}$, $Ru_3(CO)_{11}((2\text{-pentenyl})PPh_2)$, $[Ru(CO)_2(MeCO_2)_2]_n$, $[Ru(CO)_4(\mu\text{-}MeCO_2)(CH_3CN)_2]$, $[Ru_2(CO)_4(\mu\text{-}MeCO_2)_2(\mu\text{-dppm})_2PF_6]$	Linear α-olefins	Could isomerize 1-pentene into 2-petene in benzene solution at 80°C	[34–38]
	$Os_3(CO)_{12}$, $Co_2(CO)_8$, $Mo(CO)_6$		Could isomerize 1-pentene into 2-petene in benzene solution at 80°C	
	$CoCl_2$, Grignard reagents with ligands			
	$Fe(CO)_n$(n=5,9,12 ...)	Vinylene olefin	Forms di-substituted internal olefin under 150~350°C with a minor amount of tri-substituted internal olefin	[61]
	$Fe_3(CO)_{12}$	3-ethyl-1-pentene	Forms 3-ethyl-2-pentene at 25°C, 12 hrs	[62]
	$RhCl_3H_2O$/Methanol/Ethanol; $RhCl_3/OH^-$;$[Rh(OH)\text{-}(COD)]_2$	Propenyl aromatic compounds	A very original application study of using Rhodium in C=C migration	[34, 61, 62]
	$Ir(H)_2X\text{-}(PtBu_2Ph)_2$, X=F, Cl, Br, I, OH or OCH_2CF_3, polymer carriered irdium catalysis system; R-$[(PPh_2)_2Ir\text{-}(H)_2THF_2]^+PF_6^-$		Propenyl C=C migration, intermediates of propenylbenzene with different subsitituded functional group are essential materials for perfume, antibiotics, pesticides, etc.	[54, 55]
	$RhCl_33H_2O$/Ethanol, $RhCl_33H_2O/PF_6$-Methanol; $RuCl_2(PPh_3)_2$, carried by sol–gel method			[56–58]
	$Na_2Fe(CO)_4$/CuCl or $BrCH_2CH_2Br$			[63]
	$HCo(CO)_4$			[64]

Method	Catalysis System	Olefin and Derivatives	System Features	References
Photo Catalysis	Metal Cluster Compound	Internal Olefin (C_4~C_{20})	(CnRn)aFebMc(CO)dLe, M=Mo, Ru, Rh, Sc, Ti, Cr, Mn, Co, Ni, Cu, Zn. L=V group ligands. Hihg er	[65]
	Pd@TiO_2	Linear α-olefins	Day light, selective isomerism into E isomer.	[66]
	$Fe(CO)_5$		Room temperature, day light, isomerizing 1-propenylbenzene into 2-propenylbenzene, good selectivity	[67]
	$[HFeM(CO)_8L^-]$(M=Cr, Mo; L = CO)		Room temperature, day light, THF solution system, isomerizing 1-propenylbenzene into 2-propenylbenzene, conversion rate >98%, E/Z ratio = 10:1	
	Supported heteropolyacid catalyst	Linear α-olefins	Room temperature and pressure, visible light, fully conversion into internal olefin	[68]
	Supported heteropolyacid catalyst	Oleic acid	Room temperature and pressure, visible light, fully conversion into Ω-3, Ω-6	

Table 3.
Catalytic system for olefin isomerization.

Acknowledgements

The work was supported by Beijing Science and Technology Project (code: Z201100008420019), the National natural science foundation of China youth program (NSFC code 21808241) and Fund of the Director of Technical Institute of Physics and Chemistry, Chinese Academy of Sciences. Many thanks are given to our colleagues in the Technical Institute of Physics and Chemistry, CAS, as well as the professionals in other institutes for their great help in the complementary and publication of this work.

Conflict of interest

The authors declare no conflict of interest.

Author details

Wangjing Ma[1*], Bonan Liu[2], Duanda Wang[1], Jun Zhao[1], Lu Zhang[1] and Lei Zhang[1]

1 Technical Institute of Physics and Chemistry, Chinese Academy of Sciences, Beijing, China

2 State Key Laboratory of Heavy Oil Processing, China University of Petroleum (Beijing), Beijing, China

*Address all correspondence to: wjma@mail.ipc.ac.cn

References

[1] Higman C S, Lanterna A E, Marin L, Scaiano J C, Fogg D E. Catalyst Decomposition during Olefin Metathesis Yields Isomerization-Active Ruthenium Nanoparticles [J]. Chemical Catalyst and Chemistry, 2016, 8(15): 2446-2449.

[2] Ipatieff, Pines, and Schaad, J. Am. Chem. Soc. 56, 2696 (1934).

[3] Asinger F, Fell B, Collin G. Über die doppelbindungsisomerisierung bei höhermolekularen olefinen, IV. Über den bindungsisomerisierenden einfluß verschiedener verbindungen auf n-Undecene [J]. Chemische Berichte, 1963, 96(3): 716-735.

[4] Krzywfcki, M. Marczewski, R. Modzelewski, K. Pelszik and St. Malinowski, Superacid Alumina Catalysts II [J]. React. Kinet. Catal. Lett., 1980, 13(1): 1-6.

[5] Modhera B, Chakraborty M, Parikh P A., 1-Hexene Isomerization Over Nano-crystalline Zeolite Beta: Effects of Metal and Carrier Gases on Catalytic Performance [J]. Catalysis Letters, 2009, 132(1):168-173.

[6] Shi G, Shen J. Skeletal Isomerization of 1-Hexene over Sulfided Co/Co-MCM-41 Catalysts [J]. Energy and Fuels, 2009, 23(1):320-326.

[7] Rao Y, Junjie Kang A, Antonelli D. 1-Hexene Isomerization over Sulfated Mesoporous Ta Oxide: The Effects of Active Site and Confinement [J]. Journal of the American Chemical Society, 2008, 130(2):394-395.

[8] Rao Y, Kang J, Trudeau M., Investigation of the catalytic activities of sulfated mesoporous Ti, Nb, and Ta oxides in 1-hexene isomerization [J]. Journal of Catalysis, 2009, 266(1):1-8.

[9] T. Okuhara, N. Mizuno, M. Misono. Catalytic chemistry of heteropoly compounds [J]. Advances in Catalysis, 1996, 41(8), 113-252.

[10] Bruno J E, Dooley K M. Double-bond isomerization of hexadecenes with solid acid catalysts [J]. Applied Catalysis A General, 2015, 497:176-183.

[11] C.L Hill (Ed.). Special Thematic Issue on Polyoxometalates [J]. Chemical Reviews. 1998, 98(1):388.

[12] Du Z. P. Double bond migration of 1-hexene and 1-octene catalyzed by phosphotungstic acid [J]. China Surfactant Detergent and Cosmetics. 1996(1): 12-14.

[13] Uematsu T., Tsukada K., Fujishima M., et al. The isomerization of 1-butene over cation-exchanged acidic resin [J]. Chemischer Informationsdienst, 1974, 5 (23):369–375.

[14] Gee J C, Prampin D S. A kinetic and mechanistic study of the double bond and skeletal isomerization of 1-tetradecene on SAPO-11[J]. Applied Catalysis A General, 2009, 360(1): 71-80.

[15] Harmer M A, Sun Q. Solid acid catalysis using ion-exchange resins [J]. Applied Catalysis A General, 2001, 221 (1–2):45-62.

[16] Sun Q, Harmer MA, Farneth WE. But-1-ene isomerization over Nafion(R) resin/silica composite catalyst [J]. Chemical Communications, 1996, 10 (10):1201-1202.

[17] Slomkiewicz P M. Isomerization of C-4 alkenes catalyzed by macroporous ion exchange resin [J]. Reactive and Functional Polymers, 1997, 33(2): 299-304.

[18] Ghosh R, Bandyopadhyay A R, Jasra R. Mechanistic Study of the Oligomerization of Olefins [J].

Industrial and Engineering Chemistry Research, 2014, 53(18):47-55.

[19] Onopchenko A, Cupples B L, Kresge A N. Boron fluoride-catalyzed oligomerization of alkenes: structures, mechanisms, and properties [J]. Industrial & Engineering Chemistry Product Research and Development, 1983, 22(2):182-191.

[20] Höchtl M, Jentys A, Vinek H. Isomerization of 1-pentene over SAPO, CoAPO (AEL, AFI) molecular sieves and HZSM-5[J]. Applied Catalysis A General, 2001, 207(1–2):397-405.

[21] Ahn J H, Ihm S K, Park K S. The effects of the local concentration and distribution of sulfonic acid groups on 1-butene isomerization catalyzed by macro porous ion-exchange resin catalysts [J]. Journal of Catalysis, 1988, 113(2):434-443.

[22] Petrus L, Roo R W D, Stamhuis E J, Joosten. G.E.H. Kinetics and equilibria of the hydration of propene over a strong acid ion exchange resin as catalyst [J]. Chemical Engineering Science, 1984, 39(3):433-446.

[23] Slomkiewicz P M. Isomerization of C-4 alkenes catalyzed by macro porous ion exchange resin [J]. Reactive and Functional Polymers, 1997, 33(2): 299-304.

[24] Gee J C, Williams S T. Dimerization of linear olefins on Amberlyst®15: Effects of chain length and double-bond position [J]. Journal of Catalysis, 2013, 303:1-8.

[25] Ozaki A, Tsuchija S. The isomerization of n-butenes over a deuterated ion exchange resin [J]. Journal of Catalysis, 1966, 5(3): 537–539.

[26] Kalló D, Preszler I. n-Butene isomerization on acidic ion-exchange resin [J]. Journal of Catalysis, 1968, 12 (1):1-4.

[27] Manassen J, Khalif S. Organic polymers: Correlation between their structure and catalytic activity in heterogeneous systems: III. Acid-type catalysis, sulfonated and phosphonated polyphony [J]. Journal of Catalysis, 1967, 7(2):110-116.

[28] Manassen J, Klein F S. 822. Reactions of n-butene and butan-2-ol in dilute acid. The elucidation of the mechanism and the intermediate in elimination from secondary alcohols and in the hydration of olefins [J]. Journal of the Chemical Society, 1960, 3 (1):4203-4213.

[29] Takayoshi U. The Isomerization of n-Butenes over Acidic Ion-exchange Resin [J]. Bull. chem. soc. jpn, 1972, 45: 3329-3333.

[30] Lutz E F. Shell Higher Olefins Process [J]. Journal of Chemical Education, 1986, 63: 202.

[31] Wolff W F. (Standard Oil Co., USA). Isomerization of terminal olefins [P].US Patent 3405196, 1968.

[32] Farley F F. (Shell Oil, USA). Alpha-olefin production [P].US Patent 3647906, 1972.

[33] Parshall, G. W.; Ittel, S. D. Homogeneous Catalysis: The Applications and Chemistry of Catalysis by Soluble Transition Metal Complexes, 2nd ed.; Wiley-VCH: Weinheim, 1992.

[34] Sharma S K, Srivastava V K, Jasra R V. Selective double bond isomerization of allyl phenyl ethers catalyzed by ruthenium metal complexes [J]. Journal of Molecular Catalysis A Chemical, 2006, 245(1):200-209.

[35] Fritz H E, Atkins K E, O'Connor G L. Corporation (United Carbide Co., USA), Production of Alkylidenebicycloheptenes [P]. US Patent 3347944, 1967.

[36] Nagase T, Suzukamo G, Fukao M, Nishio K. (Sumitomo Chemical Co., Ltd, USA). Production of alkylidenebicycloheptenes [P]. US Patent, 3897509, 1975.

[37] Puriwat J, Chaitree W, Suriye K, Dokjampa S, Praserthdam P. Elucidation of the basicity dependence of 1-butene isomerization on $MgO/Mg(OH)_2$ catalysts [J]. Catalysis Communications, 2010, 12(2):80-85.

[38] Hassam M, Taher A, Arnott G E, Green I R, van Otterlo W A L. Isomerization of Allylbenzenes [J]. Chemical Reviews, 2015, 115(11): 5462-5569.

[39] Jr L S G, Mcdonough J M. Selective isomerization of 1-olefins to 2-olefins [P]. US Patent, 3697616, 1972.

[40] Hu S, Liu D, Li L, Guo Z, Chen Y, Borgna A. Highly selective 1-heptene isomerization over vanadium grafted mesoporous molecular sieve catalysts [J]. Chemical Engineering Journal, 2010, 165(3):916-923.

[41] Gajda G J. Butene isomerization process [P]. US Patent, 5132484, 1992.

[42] Gajda G J, Barger P T. Olefin isomerization process [P]. US Patent, 5463161, 1995.

[43] Gajda G J, Barger P T. Olefin isomerization process [P]. US Patent, 5191146A.1993.

[44] Gajda G J, Barger P T. Pentene isomerization and etherification [P]. US Patent, 5292984, 1994.

[45] Gajda G J, Barger P T, Abrevaya H. Olefin isomerization process [P]. US Patent, 5336831, 1994.

[46] Lawson R J, Richmond D M, Gajda G J, Barger P T, Pentene isomerization process [P]. US Patent, 5367101, 1994.

[47] Lok B M, Messina C A, Patton R L, Gajek R T, Cannan, T R, Flanigen, E M. Crystalline silicoaluminophosphates [P]. US Patent, 4440871, 1984

[48] Messina C A, Lok B M, Flanigen E M, Crystalline ferroaluminophosphates [P]. US Patent, 4554143, 1985.

[49] Wilson S T, Flanigen E M. Crystalline metal aluminophosphates [P]. US Patent, 4567029, 1986.

[50] Lok B M T, Marcus B K, Flanigen E M. Manganese-aluminum-phosphorus-silicon-oxide molecular sieves [P]. US Patent, 4793833, 1988.

[51] Föttinger K, Kinger G, Vinek H. 1-Pentene Isomerization over Zeolites Studied by in situ, IR Spectroscopy [J]. Catalysis Letters, 2003, 85(1):117-122.

[52] Föttinger K, Kinger G, Vinek H. 1-Pentene isomerization over FER and BEA [J]. Applied Catalysis A General, 2003, 249(2):205-212.

[53] Müller H, Koehl H, Pommer H. (BASF, USA) Production of 2-Methyl-2-hepten-6-one [P]. US Patent, 3670028, 1972.

[54] Cerveny L, Krejeikova A, Marhoul A, Ruzicka V. Isomerization of Eugenol to Isoeugenol [J]. Reaction Kinetics and Catalysis Letters, 1987, 33 (2): 471–476.

[55] Clive D L J, Sannigrahi M, Hisaindee S. Synthesis of (±)-Puraquinonic Acid: An Inducer of Cell Differentiation [J]. The Journal of Organic Chemistry, 2001, 66(3): 954–961.

[56] Fanun M, Ayad Z, Mudalal S, Dahoah S, Meltzer D, Schwarze M, Schomacker R, Blum J. Characterization of Water/sucrose Laurate/n-propanol/allylbenzene Microemulsions [J]. Journal of Surfactants and Detergents, 2012, 15(4): 505–512.

[57] Cooper A C, Caulton K G. Identification of an Elusive Catalyst: IrH (η2-C6H4PtBu2) (Cl) (PtBu2Ph) as a Precursor for C=C Bond Migration [J]. Inorganica Chimica Acta, 1996, 251(s 1-2): 41–51.

[58] Baxendale I R, Lee A L, Ley S V. A Concise Synthesis of Carpanone Using Solid-Supported Reagents and Scavengers [J]. Journal of the American Chemical Society, Perkin Transactions 1, 2002, 1850 1857.

[59] Castonguay A, Brassard P. C-Alkylation of 1,3-Dihydroxyanthraquinones. Total Syntheses of (±)-Averufin and (±)-Bipolarin [J]. Canadian Journal of Chemistry, 1977, 55(8): 1324–1332.

[60] Shao M B, Liu G Y, Zhao J J, Wang X Y, Wang J H. $RuCl_3 \cdot 3H_2O$ Mediated Olefin Isomerizations in Ionic Liquids: A Highly Recyclable System for Olefin Isomerizations [J]. Chemical Research in Chinese Universities, 2012, 28(1): 67–69.

[61] Peters B C. Olefin isomerization process [P]. US Patent, 4587374, 1986.

[62] Casey C P, Cyr C R. Iron Carbonyl Catalyzed Isomerization of 3-Ethyl-1-pentene. Multiple Olefin Isomerizations via a π-Allyl Metal Hydride Intermediate [J]. Journal of the American Chemical Society, 1973, 95(7): 2248-2253.

[63] Reddy M R, Periasamy M. Isomerization of 1-Alkenes Using the Na2Fe (Co) 4/CuCl and Na2Fe (Co) 4/BrCH2CH2Br Reagent Systems [J]. Journal of Organometallic Chemistry, 1995, 491: 263-266.

[64] Roos L, Orchin M. Allylbenzene Isomerization Catalysed by Deuteriocobalt Tetracarbonyl [J]. Journal of the American Chemical Society, 1965, 87(23): 5502–5504.

[65] Cirjak L M, Sutherland L. Process for the photocatalytic isomerization of olefins utilizing a mixed metal cluster [P]. USPatent, 4504371. 1985.

[66] Elhage A, Lanterna A E, Scaiano J C. Tunable Photocatalytic Activity of Palladium-Decorated TiO2: Non-Hydrogen-Mediated Hydrogenation or Isomerization of Benzyl-Substituted Alkenes [J]. American Chemical Society Catalysis, 2017, 7(1): 250-255.

[67] Tooley P A, Arndt L W, Darensbourg M Y. Olefin Isomerization Catalysis by Heterobimetallic Hydrides, HFeM $(CO)_8L^-$(M = Cr, Mo, W; L = Co, Pr)[J]. Journal of the American Chemical Society, 1985, 107(8): 2422-2427.

[68] Xiao T C, Ma W J. Method of Double-Bond Migrate of Terminal Olefins [P]. GB Patent, 053532, 2016.

Section 4

Polyolefins

Chapter 5

Extending Alkenes' Value Chain to Functionalized Polyolefins

Zahra Balzade, Farhad Sharif
and Seyed Reza Ghaffarian Anbaran

Abstract

Naphtha is one of the crude oil distillation products, bringing almost the lowest value-addition to crude oil, compared to other refinery products such as liquid petroleum gas, gasoline, and diesel. However, Naphtha can be converted to one of the highest value products at the end of the value chain, i.e., polyolefins. Although the production of conventional commodity polyolefins from crude oil, is considered as one of the final products in alkenes' value chain, there are specialty polyolefins with higher values. Specialty polyolefins are small volume, high-performance thermoplastics with high-profit margins compared to traditional commodity polyolefins. Recently, some special purpose functionalized polyolefins have been developed as efficient substituents for high-performance engineering thermoplastics. Polyolefins are exploited as cost-effective platforms to produce these functionalized thermoplastics. They are promising candidates for replacing high-performance polymers with high-cost raw materials and elaborate production processes. So, functional polyolefins have introduced a new paradigm in the production of high-performance thermoplastics, extending the alkenes' value chain and increasing profitability. High-performance specialty polyolefins may find exceptional markets in niche applications. In this chapter, the commercial specialty and functional polyolefins' current situation and prospects are reviewed.

Keywords: Alkenes' value chain, specialty polyolefins, niche markets, functional polyolefins, single-site catalysts, insertion polymerization

1. Introduction

The value chain notion refers to all value-creating activities, potential developments, and opportunities in industrial production, from the raw materials to the goods delivered to the final consumers. It is an appropriate base for the creation of strategic niches. The concept of niches corresponds to innovative studies and is used in different ways. Commonly, it is related to the emerging new technology-based markets. New strategic niches are the main origins of pioneering innovations. The path-breaking works deviate from the dominant routes because of their production technology and markets. Although strategic niches may lack well-defined market in their emerging phase, some actors supply the essential resources with high expectations about the technology economy and sustainability.

The sustainability transition and radical shifts toward the value-added products in the petrochemical industry are hindered by difficulties in the production

IntechOpen

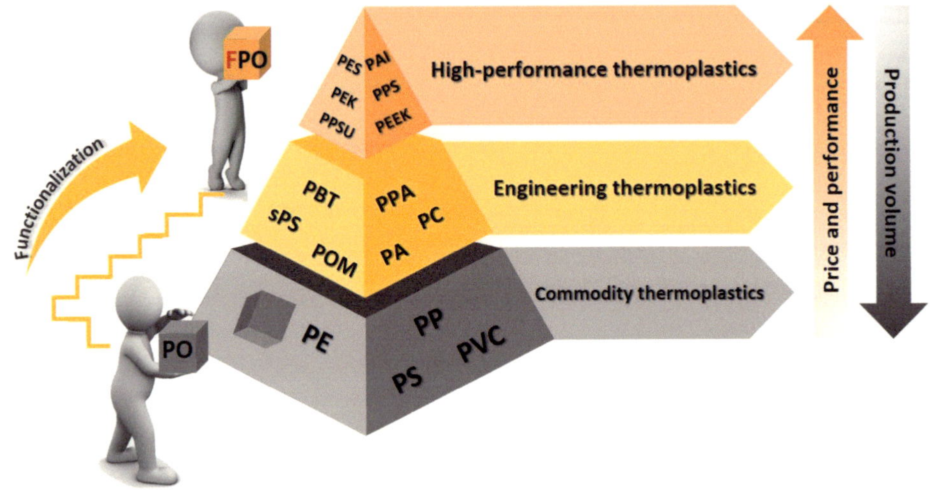

Figure 1.
Olefin-based functional co- or terpolymers as cost-effective high-quality substituents for high-performance thermoplastics.

trajectory due to the lack of well-established value chains. Besides the separated links, the lack of cross-chain interactions and connections between the suppliers and end-users hamper the petrochemical industry performance, too.

This chapter aims to evaluate how specialty polyolefins may accelerate innovative and sustainable research in industrial and/or academic areas. **Figure 1** demonstrates the thermoplastics triangle and the state of functional polyolefins, to enlighten the techno-scientific importance of emerging new products and their added value.

2. Alkenes' value chain

The alkenes' value chain starts with Naphtha or ethane (**Figure 2**). Alkenes are mainly converted to polyolefins (e.g., up to 60 wt% of ethylene converted to

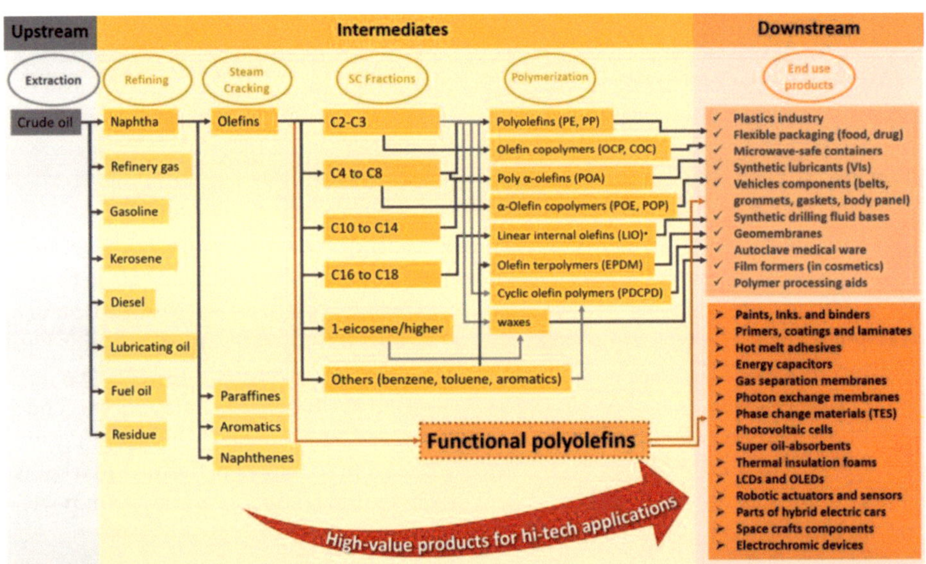

Figure 2.
*Alkenes' value chain. * Isomerization product.*

polyethylene), valued to have the largest market with more than 50 wt% of the synthetic polymers and 300 current commercial grades [1, 2]. In the downstream, polyolefin grades are converted to end-products.

The present study is focused on specialty polyolefins production, applications, challenges and future trends. Specifically, we want to answer the following questions. a) what is the potential value of specialty polyolefins? b) what are the essential requirements in taking full advantage of these value-added polyolefins? c) what are the trends and elements of the future of functional polyolefins?

3. Polyolefins' "Golden age"

Nowadays, polyethylene and polypropylene, as the most popular, widespread, and predominantly produced commodity polymers, account for approximately 32 and 27% of the global polymer market, with an annual growth rate of 4.0% and 5.3% within the forecast period of 2019 to 2027, respectively [3]. With a global market value of $300 billion and volume of 180 million tons they play an important role in our daily life [4].

The "Golden age" of polyolefins started with the seminal discovery of the Nobel Prize winners, Prof. Karl Ziegler and Prof. Giulio Natta, which was followed by finding the magical role of methyl aluminoxane in the activation of single-site transition metal catalysts, by Prof. Walter Kaminsky and Prof. Hansjörg Sinn. Then, the late transition metal catalysts bearing α-diamine ligands were introduced by Prof. Maurice Brookhart. These catalysts benefited by their exceptionally high activity, besides high tolerance toward heteroatom-containing moieties, which genuinely elevates polyolefins value to a higher level than the corresponding traditional grades [5–7].

Standing on the shoulders of the giants, the past 70 years of polyolefins technology progress has been summed up in the discovery of new catalysts accompanied with the development of new industrial polymerization processes [8, 9]. As low-cost thermoplastics, polyolefins gain the attention of researchers as an abundant platform for the production of specialty and functional polyolefins [10, 11]. The worldwide interest in specialty polyolefins and research outputs by regions over the last two decades is presented in **Figure 3**.

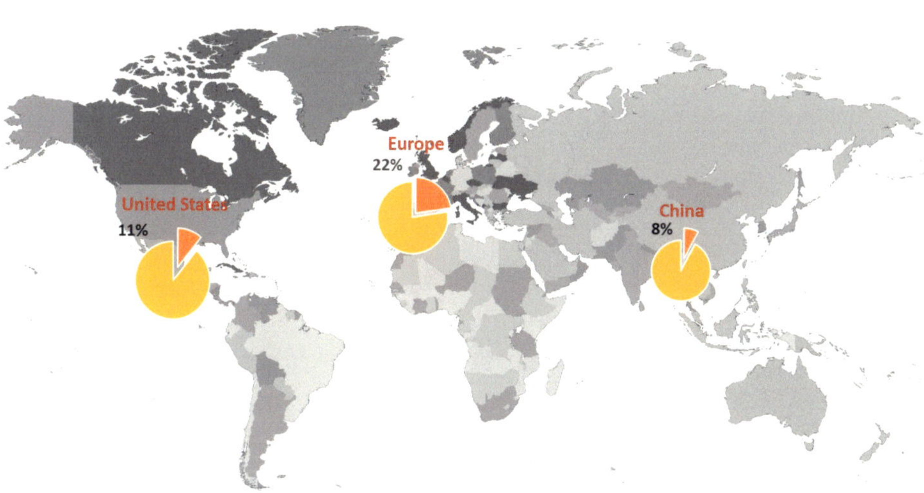

Figure 3.
Worldwide scientific interest in functional and specialty polyolefins in the last two decades.

4. Applications and niche markets of specialty polyolefins

Until 30 years ago, the physical properties of polyethylene such as density and melt index values as well as chemical characteristics i.e., inherent hydrophobicity were limited. However, the innovative single-site catalysts have removed the barriers and polyethylene products with lower densities became available. Generally, such polyethylene grades are ethylene-α-olefin copolymers where α-olefin content exceeds 40 wt% in some cases and results in elastomeric properties. In addition, ethylene-cyclic olefin (e.g., norbornene or cyclopentadiene) copolymers, also known as specialty polyolefins, extend the polyolefin properties boundaries and the alkenes' value chain.

On the other side, according to their non-polar nature, commercial polyolefins suffer from a lack of adhesion, printability, wettability, reactivity, compatibility, and miscibility. Hence, applying such an intrinsically hydrophobic polymer as an adhesive, coating, paint, primer, or binder is restricted. Also, the poor interaction of polyolefins with other materials like paints, pigments, glass fibers, clay, carbon black, metals, and the vast majority of polymers limited their applications in blends and composites [5].

However, incorporating functional groups on the polyolefin chain, even at the chain-ends, branch-ends or backbone, leads to the enhancement of polyolefins' adhesion and binding properties, which opens up exciting industrial application fields. So far, functional polyolefins' applications comprised of binders in book-binding, glue in adhesives, printing inks, tie-layer in multilayer sheets or films, and compatibilizer in blends of polar and non-polar polymers. Interestingly, they may open up new possibilities for application in film capacitors with superior energy density, proton exchange membranes for fuel cells and super-absorbents suitable for the oil spills (without any water) absorption [12].

Herein, commercially available specialty polyolefins, including functional olefin copolymers as valuable extenders of the alkenes' value chain, are comprehensively reviewed. Some of the structures of value-added polyolefins presented in this chapter are shown in **Figure 4**.

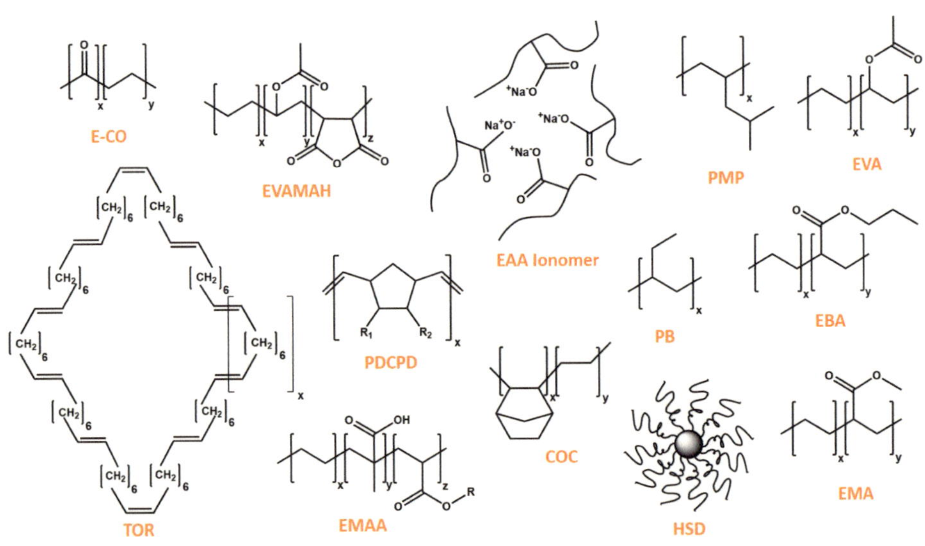

Figure 4.
The chemical structure of some of the specialty polyolefins with niche applications which will be discussed in this chapter.

4.1 α-Olefin (co)polymers

Poly(1-butene) (PB) or polybutylene is a specialty polyolefin with niche applications. This stereospecific (isotactic), linear, high molecular weight, low density, and crystalline thermoplastic has extraordinary creep, abrasion, and environmental stress cracking resistance besides superior chemical inertness. According to its unique properties and performance, some of the niche applications of this poly α-olefin include pressurized hot- and cold-water pipes, hot-water blow-molded tanks, plumbing, and heating systems [2]. Moreover, since PB has good compatibility with most of the tackifiers, it is used in hot melt adhesive formulations, especially to increase the sticky "open time" [13]. Also, PB/PE blends are utilized in easy open flexible packaging technology.

Recently, some PB composites with high dielectric constant and break downfield strength besides low dielectric loss are reported, which are susceptible to application in capacitors [14].

Hydrogenated poly(1-butene) (HPB) is a safe polyene and can be used as an ingredient in cosmetics [15]. It is used at lower molecular weights (400–1000 Daltons) in lip gloss formulations since such polyene is very shiny and possessing good adhesion to lip surfaces. Other applications of HPB in cosmetics include eye makeup, eyeliner, blushes, and foundation [16].

Hydrogenated poly(1-decene) (HPD) is commonly used as a film former in cosmetic formulations. HPDs are available in a wide range of viscosities and are used in skincare, eye shadow, makeup, and lip products due to their non-greasy skin feel [16]. The catalytic oligomerization of 1-decene to tri-, tetra-, and penta-decene chains followed by hydrogenation, results in the production of this polyene. It is available under the tradename of PAO 4–8 by Chevron Philips. Since this polyene is of low acute toxicity and does not raise concern for genotoxicity, it can be used as an ingredient in foods, such as glazing or polishing agent for dried fruits [15, 17]. Other applications of HPD are release coating in bread and loaf pans, lubricant in automatic dough dividers, anti-dusting and anti-foaming agents, and plasticizer in food contact films [18].

Although 4-methyl-1-pentene is one of the comonomers used in linear low-density polyethylene production, **poly(4-methyl-1-pentene) (PMP)** is a commercially available specialty engineering polyolefin under the trade name of TPX™. PMP has extremely low density (0.835 g/ml) and surface energy, excellent heat resistance, and chemical inertness besides high transparency, optical and acoustic properties. The industrial applications of PMP include mandrels and sheaths in high-pressure rubber hose fabrication, release films and papers in synthetic leather production, and LED light mold cups. The extremely low density of PMP makes it as a good candidate for automotive parts due to its lower weight and environmental load. Moreover, it is used in several food-related articles, e.g., baking boxes, food packs and wraps, and microwave-safe dishes [2].

Also, it is suitable for applications where transparency and heat resistance are required, e.g., autoclave medical and laboratory ware, microwave and oven components, and heat resistant wire and cables. Furthermore, since PMP is incompatible with some of the thermoplastics (e.g., poly (ethylene terephthalate)), it is used to create porous PET structures [19].

On the other side, PMP is commercially available as hollow fibers (HFs) used in gas separation membranes, where the gas permeability derived from its molecular microstructure. As it was reported recently, the transport and separation performance of these HFs improved by PMP's crystallinity increment [20].

Ring-opening metathesis polymerization (ROMP) of cyclooctene gives **polyoctenamers** which are marketed under the name of Vestenamer. They are the world's most versatile rubber additive used for years. As a semicrystalline, double bond containing polyolefin additive, trans-polyoctenamers (TORs) act as plasticizers in rubber compounding and processing. The cis/trans ratio of polyoctenamers, which is controlled during the polymerization, determines the degree of crystallinity. Generally, the higher the trans content, the higher the crystallinity and melting point. Also, the extremely high crystallization rate of polyoctenamers is advantageous in the soft compounds cold flow reduction, improving green strength, or reducing shrinkage during calendaring.

Since polyoctenamers take part in the vulcanization process, one may consider them as reactive plasticizers. Also, the high macrocycle content of polyoctenamers reduces their molecular weight significantly. In addition to the low molecular weight, the broad molecular weight distribution of polyoctenamers is responsible for their unusual low viscosity at elevated temperatures. Polyoctenamers enable the compounding of incompatible polymers (e.g., NBR and EPDM), reduce viscosity, heat build-up, mix time and energy intake, improve dimensional stability, quality and filler acceptance, and increase throughput.

Furthermore, polyoctenamers allow efficient processing of waste rubbers into tough materials to be used again, i.e., effective in rubber sustainability [21].

Vistamaxx is the trade name of **propylene-ethylene copolymer (PEC)** produced using metallocene catalysts by ExxonMobil Chemicals, comprised of isotactic propylene blocks with random ethylene distribution where the propylene content is over 70 mol%. This copolymer's chemical structure and properties are intermediate between amorphous ethylene-propylene rubber and semi-crystalline isotactic polypropylene [22]. Considering PEC's good processability and compatibility with a wide range of polyolefins, and very low viscosity, it is used as an ingredient in hot melt adhesives (accounts for up to 70–90% of the mixture providing high-performance HMAs with tunable "open-time") and as a processing aid or viscosity improver in extrusion or injection molding processes to improve the flow characteristics.

Consequently, PEC enhances polyolefin blend properties to deliver improved impact strength, higher flexibility, enhanced esthetics, and lower stress-whitening. Also, it is employed as a sealing layer in co-extruded articles due to its extremely low seal initiation temperature combined with high seal strength. This copolymer is a suitable substituent for wax in masterbatch formulations, optimizing costs by lowering the cycle time. Moreover, the color strength increased in the presence of such copolymer by almost 20% because of its high compatibility with polyethylene and polypropylene and well pigment wetting and dispersion at lower processing temperatures [23–25].

Moreover, using Vistamaxx as a minor component (up to 20 wt%) in spunmelt nonwovens designed for applications such as leg barrier cuffs, diapers and medical gowns results in improved softness [26].

Olefin copolymer (OCP) viscosity improvers (VIs) are mainly ethylene-propylene copolymers with ethylene to propylene ratio of 45/55–55/45 synthesized using Ziegler-Natta and metallocene catalysts. Short ethylene sequences in OCPs, are crystallized into fringed micelles while higher-ordered morphologies, i.e., lamellae or spherulites, are absent in such copolymers. Also, the fraction of EE sequences (dyads) and the average length of contiguous ethylene units are raised by increasing the ethylene to propylene ratio. Due to the limited solubility of such OCPs in most mineral oils and consequently the inappropriate function of OCP as VI, the degree of crystallinity should not exceed 25% in OCP VIs [27]. Therefore, the physical properties of OCPs fall in between the characteristics of semicrystalline polyethylene and amorphous ethylene-propylene rubber. This oil-soluble copolymer cold flows at room temperature and improves the low-temperature rheology, thickening efficiency, and bulk handling characteristics of engine oils.

Moreover, some essential functions such as dispersing contaminants, antioxidative stabilization, and antiwearing are combined with the rheology control features on the same molecule by OCPs functionalization. Some hybrids have been commercialized, including dispersant OCPs, dispersant antioxidant OCPs, and ashless antiwear dispersant OCPs. These hybrid lubricants are produced through different methods. The most attractive approaches are free radical grafting of nitrogen-containing monomers such as phenothiazine to provide antioxidant functionality and two-step grafting method. In the second approach, maleic anhydride (or such diacyls) is grafted onto the OCP molecule. Then the amine or alcohol reacts with the anhydride to create imide, amide or ester bonds. Also, many functional OCPs have been described in the patents [28–30].

The future perspective of OCPs includes optimizing viscosity ranges, improving the shear stability of the formulations, and designing novel tailor-made functional OCPs (FOCPs) as low cost, highly efficient, and customer choice multifunctional VIs [27].

Ethylene-propylene-diene terpolymers (EPDMs) industrial production returns to the 1960s by ExxonMobil Chemical Company, which was designated as Vistalon™. The development of this terpolymer was a natural evolution of Ziegler-Natta technology after the production of other polyolefins. The saturated backbone of EPDM, compared to other rubbers, and consequently its excellent ozone, environmental, and weather resistance makes it the material of choice in the production of various outdoor articles. Typical applications of EPDM include the automotive industry (weather seals, radiator hose, gaskets, grommets, O-rings, belts), construction industry (single-ply roofing material, geomembranes), electrical devices (wire and cable covers), polymer modification (polymer blends used in automotive bumpers), oil modifiers and so on [31, 32].

Based on developments in Daw Chemicals catalyst technology, novel grades of EPDM, i.e., ultrahigh molecular weight EPDM, known as (UHMW)NORDEL™ were developed. NORDEL is highly efficient in producing thermoplastic vulcanizates (TPVs), light color home appliance gaskets, and black color low hardness molded articles [33, 34].

Recently, the potential future applications of novel paraffin-filled EPDM foams as phase change materials (PCMs) in thermal energy storage (TES) systems have been investigated [35, 36].

4.2 Cyclic olefin (co)polymers

As a successful example, **ethylene-norbornene copolymers,** well-known as COCs, are synthesized via the copolymerization of ethylene with norbornene (or cyclopentene) using metallocene complexes. COCs are glossy, transparent, rigid and amorphous thermoplastics commercially available by TOPAS Advanced Polymers, Inc. The low shrinkage and high modulus (due to the high norbornene content of about 30–60 mol%) of this copolymer make it available as extruded sheets, casts or blown films, and injection or blow molded finished articles. It is widely used in several applications, including medical devices, packaging films, cosmetics containers, and microfluidics. Moreover, this copolymer's high glass transition temperature (up to 180°C) and high heat distortion temperature (about 170°C) make it appropriate low-cost substituent for polycarbonates (PCs) [2].

Also, the effective thermal radiation shielding of poly (ethylene-*co*-norbornene) based COC foams may open new perspectives in the fabrication of thermal insulation foams [37].

Hydrogenated polynorbornenes as cyclic olefin polymers (COPs) were commercialized since 1990's and have found increasing applications in pharmaceutical syringes and vials thanks to their superior properties vs. glass and other transparent thermoplastics. This engineering thermoplastic was synthesized via ROMP of norbornene using Grubb's catalyst and subsequent in situ hydrogenation [38].

Low protein adsorption, outstanding moisture barrier, extremely low extractable, considerable fracture resistance, and transparency make hydrogenated polynorbornene an excellent choice for prefilled syringes and vials suitable for parenteral and lyophilized biopharmaceuticals. Also, the low fluorescence, high optical transmission, low haze, and precise mouldability of such COPs make them great selections for bio-diagnostic and life-science devices. The other advantages of this polymer include low risk of interaction with drug and excellent container closure integrity (CCI) at cryogenic temperatures (polymer/rubber thermal expansion coefficients match). However other candidates suffer from limitations in one or more ways. For instance, despite the extreme resistance of poly (vinylidene chloride) to oxygen and water, its sterilization results in HCl release, which causes compatibility issues.

On the other side, the superior dimensional and optical stability of COPs, even after prolonged exposure to high humidity and heat, are beneficial options for mobile devices and large-format TVs (LCDs and OLEDs). Also, displays with excellent viewing angles can be made from COPs, due to their stable, uniform birefringence even at sharp, oblique viewing angles from any seat [39–42].

Unlike the aforementioned cyclic olefin (co)polymers, Poly dicyclopentadiene (PDCPD) is a thermoset with potential and typical applications in transportation system components such as vehicles' body panels' (cabin roof, floor, engine bonnets, mudguards), agriculture equipments, chemical and wastewater treatment, and renewable energy production (wind turbine blades). PDCPD produced by polymerization of low viscosity dicyclopentadiene using Grubb's catalyst through a ring-opening metathesis approach. It is a successful alternative for metals and ceramics in various durable articles and heavy-duty applications [2].

Recently the degradability and recyclability of thermosets like PDCPD has been investigated. It has been shown that the incorporation of some cleavable bonds within the polymer chains, facilitates the triggered degradation of such thermoset material [43].

Hydrogenated poly(1,3-cyclopentadiene) (HPCPD) is synthesized through ROMP of DCPD, followed by the double bond hydrogenation. It is a low molecular weight polymer soluble in volatile and non-volatile hydrocarbons that imparts water-proof characteristics to the formulations and adhesion [16]. HPCPD with low risk of carcinogenicity, toxicity, and allergies is used in several applications, including cosmetics (e.g., film formers, waterproofing agents and blends combinations), creams, lotions, gels, hair, skin and sun care [44–46].

For the first time, **syndiotactic polystyrene (*s*PS)** has been developed in Idemitsu Kosan Central Research Laboratory. It was synthesized using single-site cyclopentadienyl titanium trichloride-based metallocene catalysts in the presence of methyl aluminoxane as cocatalyst [47]. Syndiotactic polystyrene is currently available as XAREC™, which is a superior environmentally friendly engineering thermoplastic. It has a high melting point (270°C) compared to its isotactic analogue (240°C). More importantly, the crystallization rate of *s*PS is 40 to 80 times higher than *i*PS at the same cooling conditions. The high heat distortion temperature, excellent chemical resistance, dielectric constant, and dissipation factor are of the primary advantages of this engineering polymer.

*s*PS is utilized in temperature resistant automotive and home appliance applications and its biaxially oriented film is used in high-temperature resistant films, e.g., ovenable packaging. However, to overcome its inherent brittleness and poor impact strength, nanocomposites of *s*PS reinforced with glass or carbon fibers, mineral fillers or elastomers are developed [48].

Owe to the excellent thermal, electrical and chemical resistance, low specific density, and being environmentally friendly, *s*PS is prospective in the fabrication of electronic components of hybrid electric vehicles [49].

Styrene-diene (butadiene or isoprene) copolymers are synthesized via anionic polymerization, usually using butyllithium. **Hydrogenated styrene-diene (HSD) copolymers** are used as VI and classified into linear block and star copolymers. HSD block copolymers are synthesized via a step-by-step approach. The blocks synthesized alternatively, i.e., the first block is synthesized and then the second block is added to the "living" polymer. Star-shaped HSDs are also synthesized in two steps. Generally, the core compound is one of divinylbenzene, polyhalogenated hydrocarbons, cyclosiloxanes, or calixarenes. The arms are di- or triblock copolymers of styrene, isoprene, or butadiene, which are grown from the reactive sites on the core [50].

The amorphous nature of HSDs affects the low-temperature flow characteristics. Such di-block copolymers and the associated micelles are efficiently used to improve the oil thickness at higher temperatures (100–150°C). HSDs are mainly utilized in high-performance motor oils for gasoline and diesel engines. Moreover, HSDs will find new applications in top-tier niches thanks to the ease of design of such block copolymers with specific topologies to have specialized features [27].

4.3 Functional polyolefins

Alternative olefin-carbon monoxide copolymers are tough semi-crystalline high-performance thermoplastics synthesized by coordination polymerization using late transition metal catalysts. This copolymer has been commercialized since the late 1990s, and now available as Poketone™ copolymer (ethylene-CO) and terpolymer (ethylene-propylene-CO) produced by Hyosung Chemicals.

Unique engineering properties such as exceptional impact and wear resistance, chemical and fuel resistance, gas barrier properties, and superior ductility over a broad temperature range are the significant advantages of this engineering thermoplastic.

Olefin-CO copolymers possess 2–3 times higher impact strength than poly amide (PA) and polybutylene terephthalate (PBT), higher hydrolysis resistance than PA, and better wear resistance than polyoxymethylene (POM). This thermoplastic is applicable in reinforced thermoplastic pipes (RTP) due to its high gas and hydrocarbon barrier properties. Moreover, its considerable resistance to automotive fluids made it a good candidate for application in the fuel system. Also, since this thermoplastic is safe with low to zero volatile compounds emission, it is used in food contact packaging, toys, and medical devices [51].

Generally, since the microstructural features of olefin-CO copolymers are translated into their macroscopic properties, a detailed understanding of the polymerization mechanism and structure-properties relationships is the key to design and synthesis new olefin-CO copolymers for specific applications. According to the literature, it seems that functional olefin-CO copolymers may find considerable attention in the future. The keto groups in olefin-CO copolymers may act as chemical "hooks" for anchoring the functional groups or crosslinking and curing the olefin-CO article after the processing [52].

The other perspective is incorporating a low amount of in-chain keto groups to render photodegradable polyethylene as an environmentally friendly sustainable thermoplastic while retaining its characteristic properties [53, 54].

Ethylene-Silicone block copolymers are novel block polymers in which ethylene and silicone units are covalently bonded together, using an appropriate catalyst system.

In 2018, the Chemical Society of Japan Award for Technical Development was given to Mitsui Chemicals for Exfola™ (first commercial ethylene-Si block copolymer) production and commercialization.

This functional polyolefin is an additive that combines the characteristics of polyolefin and silicone. So, the surface properties of polyolefins (i.e., release properties, water and oil repellency, coefficient of friction, abrasion resistance, and surface smoothness) are affected by such surface modifier, especially in injection molded or extruded articles, sheets, and films. In other words, this functional olefin-based copolymer changes the surface of a molded polyolefin article to silicon-specific characteristics by adding a small amount of this surface modifier during the molding process [55].

Lotryl® MA and Lotryl® BA (SK Chemicals), Optema™ and EnBA (ExxonMobil), and EMAC® and EBAC® (westlake), all are **ethylene-methyl (or butyl) acrylate copolymers (EA)** produced by high-pressure high-temperature radical polymerization, with acrylate content of up to 40 wt%. By increasing the acrylate content, the adhesion, solubility, toughness, compatibility with polar substrates, filler acceptance, and flexibility of the copolymer are enhanced. At the same time, the crystallinity, melting point, softening point, rigidity, and hardness are decreased [56].

Thanks to their outstanding compatibility with other thermoplastics and materials besides superior adhesive properties, EA copolymers are used as impact modifiers in engineering plastics (PET, PBT), base materials for filled compounds (masterbatches, wire and cables), and sealable films and layers in flexible packaging. For example, the application of this copolymer in hot melt adhesive composition, relates to its high compatibility with other polymers and raw materials such as tackifiers, waxes, and plasticizers [57].

Although EA copolymers are widely used in several applications, the harsh radical polymerization conditions (pressure range of 250–3000 bar and temperature range of 150–375°C) lead to a poor control over the microstructure, i.e., broad molecular weight and comonomer composition distributions. Therefore, the synthesis of tailor-made polar copolymers of ethylene under mild conditions, using coordination insertion polymerization and controlled radical polymerization approaches gained much attention in the last years. Well-defined functional polyolefins with controlled architectures and topologies will increase the versatility of such copolymers and opens new horizons in polyolefin niche applications [58, 59].

The introduction of acid side branches into the polyolefin substrate brings unique functionalities to **ethylene-acrylic acid copolymers (EAA)**. The acrylic acid units enhance its adhesion to polar substrates including papers, aluminum foil, metallized films, iron, steel, glass, ionomers and polar polymers (e.g., PA and poly ethylene-*co*-vinyl acetate (EVA)), significantly.

Escor™ an Nucrel™ are commercially available ethylene-acrylic acid copolymers (6–11 wt% AA), produced by ExxonMobil and Dow Chemicals, respectively.

Unlike conventional chemical primers and adhesive solutions, EAA copolymers provide excellent high-speed extrusion coating and lamination without the need to inconvenient and rate-limiting adhesive application and drying. So, EAA copolymers are cost-effective solutions for foil adhesion. Strong bonds form between EAA and the oxide layer on the aluminum that is highly resistant against mild to moderate acidic or basic environments.

Moreover, Nucrel™ AE is a specialized ethylene-methacrylic acid-acrylate terpolymer (EMAA), providing improved foil adhesion and enhanced hot tack strength [60].

Unistole™ is a liquid thermoplastic, commercially available as **hydroxyl**- or **acid-modified polyolefin**. It is utilized as non-chlorine, colorable, primer, or adhesive compatible with almost all types of paints, used in the automotive industry, medical packaging, printing materials, bonding agents, and tackifiers besides high-performance packaging. It has superior adhesion to a wide variety of polar substrates, such as Nylon 6, polyurethane, PBT, ABS, and EVA, flexibility and heat-seal properties.

Surlyn® is a smart intrinsic self-healing thermoplastic based on **partially neutralized** (with Na^+, Mg^+ or Zn^+ ion) **ethylene-(meth) acrylic acid copolymer** synthesized via high pressure free radical polymerization using tubular reactors by Dow Chemical Company.

This ionomer enables microcrack reparation under specific triggers. Accordingly it is majorly utilized in impact protection applications such as golf balls, boats, or car bumpers [61]. Moreover, its potential application in hypervelocity impact-resistant less vulnerable spacecraft protecting bumpers against space debris has been evaluated, recently [62].

In addition, this clear, adjustable, and cost-effective engineering thermoplastic resin is utilized in packaging applications due to its excellent barrier properties and resistance to oil penetration. Also, besides the polar units within the polyolefin chain, the low sealing temperature of this ionomer makes it a good choice for adhesives and tie-layer materials. Moreover, considering the superior puncture, tear, and abrasion resistance of this ionomer, its significant applications include coatings, inks, food packaging, sporting goods, and cosmetics molded containers. The excellent heat sealability and oil and grease resistance provided by EAA ionomers make composite films gain popularity in the food packaging industry and rise in demand for such films.

Since poly (ethylene-*co*-methacrylic acid) metal composites exhibit flexibilities like traditional ionic polymer metal composites (IMPCs), they are considered as promising candidates for novel soft robotic actuators and sensors utilized in finger joints, hip, knee, or segmented limbs as well as energy harvesters [63].

Electrochromic devices (ECDs), such as chromogenic windows in aircraft, automobiles and buildings, are attractive potential applications of electrochromic materials both from the academic and industrial points of view for their selective and controlled visible light and solar energy transmission. Gelatin-based electrolyte films blended with EAA copolymer have demonstrated improved coloration efficiency compared to ones prepared by the solution mixing technique. This solvent-free approach will increase the chance of gelatin-based ECDs early commercialization [64].

Ethylene-vinyl acetate copolymers (EVA) are specialty thermoplastics produced via radical polymerization and commercially available with vinyl acetate content of 15 to 45 wt% or even more [65, 66].

EVA is an intelligent choice for flexible, puncture-resistant, and low seal initiation temperature (means faster packaging speeds) food and medical packaging which is a suitable replacement for PVC in non-invasive medical tubing and bag applications. Moreover, thanks to its good flow in the heat seal process and excellent crack resistance, EVA is suitable for block cheese, cereal, snacks, and fresh meat packaging. On the other side, most of the caps, closures and lids are made from EVA due to its good gas permeability, sealability, and heat resistance [60]. Also, EVA is used in automotive seals, for example, rocker head covers, due to its long-term heat stability besides good resistance to automotive fluids [67].

Recently, the low potential induced degradation (PID) and high transparency of new qualified grades of EVA made them promising for encapsulant sheet materials in photovoltaic cells [68].

SPO	Trade Name	Manufacturer	Functionality	Applications
PB	Akoafloor Akoalit Koattro Toppyl Purell	Lyondell Basell	—	easy-open packaging, film modification, hot melt adhesives, polyolefin modification, hot-water tanks, pipe and fittings, shoe sole
PMP	TPX™	Mitsui Chemicals	—	semiconductors, wires and cables, medical packaging, food packaging, films, household
PDCPD	Metton® LMR	Metton America	—	large truck and service vehicle body panels, construction parts, shipping containers, fan shrouds, septic and water treatment equipment, engine covers and fenders, hoods, bonnets
	Telene®	Telene RIMTEC Corporation		
*s*PS	Xarec	Idemitsu	—	automotive cooling applications automotive high voltage applications, plating grade, circuit boards, LED, food contact and potable water
EPDM	Dutral	Versalis	—	"sponge" profiles for cars, weather-sealings, building profiles, wire & cables, washing machine gaskets, home appliances, automotive interiors, automotive plastics modifiers, single-ply roofing, belts, O-rings, grummets, hoses, radiator-hose, diaphragms, accumulator bladders, geo-membranes
	Nordel	Daw Chemicals		
	Esprene™	Sumitomo		
	Keltan®	Lanxess		
	KEP	Kumho Polychem		
	Vistalon™	ExxonMobil Chemical		
COC	Topas®	Topas Advanced Polymers	—	perfect parenteral packaging, wearables, drug delivery, medical devices, recycle-friendly shrink labels & films, protective packaging, films, easy tear packaging, cellophane-like twist films, cosmetics and bottles, microplates, antennas for mobile devices
	Apel™	Mitsui Chemicals		
	Arton	JSR Corporation		
COP	Zeonex® Zeonor®	Zeon Corporation	—	pre-filled drug delivery devices (syringes and cartridges), vials and bottles for long-term biological materials storage, IV and total parental nutrition (TPN) bags, bio-reactors, high-pressure injection syringes (needle-free, viscose drug and large dosage injection), autoinjectors, cells for hem analysis, lenses, prisms, sunglass-friendly touch sensors, optical films and sheets for LCD and OLED displays, optical fibers, optical lenses in digital cameras and laser beam printers

DOI: http://dx.doi.org/10.5772/intechopen.99078

SPO	Trade Name	Manufacturer	Functionality	Applications
HPCPD	Koboguard®	Kobo Products	—	cosmetics (film formers, substantivity agents, waterproofing agents, blends combinations, sunscreen agents, lubricants slip agents, dispersants), creams, lotions and gels, decorative cosmetics and makeup, hair care (shampoos, conditioners and styling), skincare (facial care, facial cleaning, bode and baby care), sun care (sun protection, after-sun and self-tanning), lipsticks, concealer, lip balm with SPF, blush, lip gloss, bronzers, highlighters, brow liner, eye shadow, lip balm, foundation, eyeliner
TOR	Vestenamer®	Evonik	—	tires (strips, apex, tread, sidewall, recycling, bladders), rubber recycling (virgin rubber blends, rubber-modified asphalt, recycling rubber goods), rubber goods (hoses, profiles, seals, rolls, drive belts, conveyor belts, cable and wires, flooring and roofing, molded goods, foamed products), medicals (rubber stoppers, thermoformable products), sports (golf balls), shoe soles, masterbatches (additives, carbon, Sulfur), thermoplastics
HPD	PAO 4 FG	Chevron Philips	—	cosmetics primary and secondary oils (as carrier and moisturizers), glazing or polishing agent for dried fruits and particular sugar confectionery (e.g., fruit gums and jellies), substituent for white mineral oil, release ('non-stick') coating in bread tins, lubricant in dough-dividing machines, anti-dusting and anti-foaming agent, plasticizer in food contact films
PEC	Vistamaxx™	ExxonMobil Chemicals	—	hot melt adhesives, sealing layer in co-extruded articles, masterbatch formulations, compound modification, stay-flat carpet tiles, breathable soft-stretch elastic nonwoven fabrics
OCP	HiTEC®	Afton Chemicals	—	VI for automotive and industrial applications (engine oils, hydraulic fluids, tractor fluids, wireline, gear oils, grease)
	Paratone	Chevron Oronite		
	Nordel IP	Dow Chemicals		
	Lubrizol	Lubrizol		
	Viscoplex	Evonik		
	Dutral Co	Versalis		
E-Si	Exfola	Mitsui Chemicals	Silicone	surface modifier of polyolefin films, sheets, containers, and packages

SPO	Trade Name	Manufacturer	Functionality	Applications
E-CO	Poketone	Hyosung Chemicals	ketone	toys, sporting goods, food contact parts (food packaging films, water purifiers, faucets), cosmetic packaging, dispensing solutions packaging, pipes (downhole pipe liners, tubes, fitting, filaments), automotive (connectors, worm wheel hubs/gears, fuel system), medical (nasal sprays, snap-fits, valves, nozzles, surgical instruments, technical disposals), gear (glove conveyors, ATM gears, Helmet parts for motors and bikes, refrigerator rotor), electronics (mobile bracket, connector housing, Watt-hour meter), Industrial (rail fastening system, caster)
EA	Emac® Ebac®	Westlake Chemical Corporation	methyl acrylate butyl acrylate	hot melt adhesives, wire and cable, masterbatches, extrusion coated structures, extruded laminates, hot melt adhesives for deep-freeze packaging, sealable layer for food packaging, films, impact modification of engineering plastics
	Lotryl® MA Lotryl® BA	SK Functional Polymer		
	Optema™ EnBA	ExxonMobil Chemicals		
	Acryft®	Sumitomo Chemicals		
EAA	Escor™	ExxonMobil Chemicals	acrylic acid	adhesives, sealants, coatings, laminates, tube containers, transparent packaging, packaging (condiments, juices, aseptic milk, personal care products, pharmaceuticals), cable shielding, metallized building panels, water piping
	Nucrel™	Dow Chemicals		
EAA ionomer	Surlyn	DuPont	neutral acrylic acid	coatings, inks, surface coating films for golf balls, food packaging, sports equipment, molded bottles of cosmetics, medical device packaging, stretch packaging, packaging of food materials such as meat, poultry, fish, seafood, cheese, edible oil, vegetables, flexible packaging formats such as sachets, pillow pouches, flow-wrap packages, anti-corrosion paints, composite cans
	ChemipearL™	Mitsi Chemicals		

DOI: http://dx.doi.org/10.5772/intechopen.99078

SPO	Trade Name	Manufacturer	Functionality	Applications
EVA	Escorene™	ExxonMobil Chemicals	vinyl acetate	gauge films, bags, pouches, carpet backing, film wrap, lamination film, sealants, sealants in meat, dairy packaging structures, footwear, wire and cable insulation, photovoltaic encapsulants and sheets, packaging (block cheese, cereal and snacks, fresh meat barrier packaging, closures, lids) films, medical packaging, footwear, hot melt adhesive, ink, road bitumen modification, heat seal agent for paper, film and aluminum foil in packaging chocolate and confectionery, anti-blocking agents, greenhouse covers, hot melt adhesives in book-binding, modification of emulsions
	Elvax™	Dow Chemicals		
	Sumitate®	Sumitomo		
	Evatane®	SK Chemicals		
	Ultrathene	Lyondell Basell		
	Levapren®	Arlanxeo		
	Chemipearl™	Mitsui Chemicals		
	Elevate®	Westlake chemicals		
EVMA	Orevac® T	SK functional polymers	vinyl acetate-maleic anhydride	flexible packaging, multilayer blown films and cast films, shrinkable films, barrier lidding films, vacuum bags, coextruded multilayer barrier and composite pipes, multilayer tubes and pipes, pipe coating, heat shrinkable sleeves, automotive industry (door panels, seats and roofing, bonding of textile substrates in seats covers and headliners), carpets, sound proofing , heat insulation materials, sports and leisure (skis, sails, shoes (textile uppers, liners)), furniture (decorative films, mattresses, seats, wall covering, bonding of foams on woven and non-woven), shrinkable bags, lids for trays, tubular films
KEE	Elvaloy™	Dow	ketone-ester	ready-to-eat meal packaging, asphalt paving, cable jacketing, roofing membranes, PVC and other plastics modification

Table 1.
Function-application review of commercially available specialty polyolefins (SPO).

Random terpolymers of ethylene-vinyl acetate-maleic anhydride (EVAMAH) are manufactured by the high-pressure radical polymerization process. They are commercially available as Orevac® T. A transparent, flexible, soft, and reactive polyolefin-based functional thermoplastic with superior adhesive properties to several polar and non-polar substrates, generally utilized as tie-layer in multilayer films, tubes, and pipes. It is used to bond diverse solid substrates such as PA, PU, and PET films, metallic foils, fiber glass fabrics, textiles, artificial leather, natural fibers, wood, and foams.

While vinyl acetate units provide softness, flexibility, and polarity to this terpolymer, maleic anhydride gives reactivity to Orevac® T, leads to adhesive properties far better than EVA copolymers [69].

Ketone-Ethylene-Ester (KEE) terpolymers are commercially available as Elvaloy ™ provided by Dow Chemicals. KEE improves the durability, flexibility and long-lasting characteristics of other resins. In addition, when combined with PET, it provides desired level of low-temperature toughness, especially in refrigerated and frozen meals.

Moreover, KEE terpolymers modify the long-term properties and load carrying capacity of asphalt as polymeric modifiers. Also, the permanent flexibility, high cutting, puncture, shrinkage, and chemical resistance of Evaloy/PVC compounds made them the most durable, fastest to install, and most accessible to repair, high-performance single-ply PVC roofing [60].

A list of the specialty polyolefins which has been reviewed in this chapter, is given in **Table 1**. The trade name, manufacturer, Functionalityality added to the polyolefin chains, and applications of such specialties are summarized.

5. Challenges in the commercialization of functional polyolefins

Since the early 2000s, the synthesis of functional polyolefins with complex architectures faced tremendous progress thanks to the development of novel insertion polymerization catalysts enabling the synthesis of polyolefins with highly controlled molecular weight, chemical composition distribution, and topology. The precise synthesis of functional polyolefins opens up new windows toward the development of engineering grades based on structure–function relationships [10].

Undoubtedly, catalytic polymerization will stand at the core of functional polyolefin production. However, the remaining related challenges are as below;

- The environmental concerns, including polyolefins accumulation in the environment, are perhaps essential challenges. Some of the potential solutions are the production of biodegradable polyolefins, increasing their recyclability, and developing new functional polyolefins as replacements for non-olefin based materials to reduce the sorting requirement during recycling. The other potential approach is the catalytic depolymerization or upcycling of polyolefins through which valuable materials are obtained.

- The residual catalyst remains in the polyolefin matrix is another concern, not only from the toxicity point of view but also due to the undesirable degradation reactions acceleration in the presence of metallic contaminants. Also it is a substantial issue that may affect the polyolefins' recyclability and limits their applications. The most probable approach to solve this problem is developing super active catalysts capable of producing special-purpose functional polyolefins, e.g., with pendant antioxidant groups to suppress the undesirable degradation reactions during the service period [10, 70, 71]

Although random functional polyolefins are obtained by applying minimum changes compared to their corresponding homo-polyolefins, they offer drastically different chemical, physical, and surface properties. Therefore, random functional polyolefins are excellent candidates for the replacement of engineering thermoplastics, opening new windows toward the progression of new engineering polyolefins. However, despite the recent developments in the synthesis of such functional polyolefins, there are still several challenges in their production include;

- The active catalytic centers of highly exophilic early transition metal catalysts are deactivated by forming strong σ-coordination with the Lewis basic moiety of heteroatom-containing polar monomer. Accordingly, the late transition metal catalysts were introduced as promising substituents because they are more tolerant toward functional groups [12].

- The other major challenge originates from the relative electron deficiency of functional monomers compared to olefins, which results in the former lose-out during the competitive coordination to the metal center or perhaps stable chelates formation. Therefore, the poor π-donating ability of polar monomers avoids the formation of heteroatom-rich functional polyolefins.
  Also, it still suffers from the limited coordination and insertion of the polar functional groups and vast imperfections in the synthesis of well-defined tailor-made functional copolymers [11, 70].
  As a solution, the design of new catalysts capable of producing high polar comonomer content functional polyolefins seems to gain more potential in the future. However, despite the tremendous developments in highly effective olefin (co)-polymerization catalysts, only a few of these catalysts have found general applications in industrial polyolefin processes. There are a number of parameters that need to be evaluated. The parameters are catalyst activity and kinetics, range of products with a specific MW and MWD, morphology, branching density and comonomer incorporation homogeneity, tolerance toward polar monomers in the production of functional olefin copolymers as well as polymerization process economics and environmental issues [4, 11].
  Also, chain-end functionalized polyolefins synthesized via chain-transfer reactions taking advantage of almost quantitative incorporation of the functional groups at the chain ends. This class of functional polyolefins is a superior starting point for the grafting from or block copolymer synthesis, where the functional fragments are created by another reaction mechanism successively.
  Furthermore, the complexity in controlling the microstructure and composition of functional copolymers encouraged the researchers toward the copolymerization of ethylene with dormant reactive comonomers. The dormant moiety transformed to functional groups through post-functionalization approaches or to initiators for graft-from polymerization of polar monomers [5].
  The other approach comprises the precise synthesis of random functional polyolefins through metathesis polymerization.
  Moreover, different synthetic pathways, such as functionalized carbenes as alternatives for the synthesis of functional polar monomers, are constantly developing [11].

- Last but not least is related to the development of functional polypropylenes. The lower reactivity of propylene compared to ethylene makes the production of functional propylene copolymers much more difficult than ethylene copolymers [72].

6. Trends in the specialty polyolefins industry

Specialty polyolefins growing in the last decade was faster than the past thanks to developing new catalytic systems and polymerization processes.

Nowadays, the possibilities in the synthesis of specialty polyolefins include;

- Homo-, co- and terpolymerization of olefins, dienes (as dormant sites for post-polymerization modifications), and polar vinyl monomers

- Stereospecific polymerization for the synthesis of all kinds of tacticities such as stereo-block, iso-block, semi-isotactic, syndiotactic, and atactic structures with tunable block lengths

- Strategies to control the topology of polyolefins, such as chain walking polymerization. Using late transition metal catalysts, a wide range of engineered polyolefins of complex chain architectures can be designed through the control of ligands' steric and electronic parameters [73–75].

The progress in specialty polyolefins production will help redefining polyolefins' intrinsic properties such as tackiness, biodegradability and sustainability [4]. Therefore, the future possibilities in specialty polyolefins production will be as follows;

- Controlled introduction of branches with tunable lengths

- Production of polyolefins with defined tacticities of prochiral monomers in the backbone

- Tailored co- and terpolymers with defined sequence and combination of monomers and topology. Examples include telechelic, comb-like, star, miktoarm, dendritic, and hyperbranched functional polyolefins or 3-D structures with self-organizing helix or zig-zag units, rigid or elastic blocks, and more sophisticated systems [76, 77].

- Multiblock polyolefins synthesized in the presence of chain shuttling agents

- Functional polyolefins obtained via hybrid insertion/radical polymerization

- In situ polyolefin nano- or micro-composites

- Polyolefin-based nano-catalysts, nanoreactors, and nanomachines taking advantage of the polyolefin conformational switch under various triggers according to the functionality and topology. The nanoreactors provide desired reaction conditions by allowing only suitable reactants to reach the embedded catalyst elements

Moreover, due to the emerging role of single-site catalysts in the development of specialty and functional polyolefins, the future trends in catalyst progression are listed as below;

- Production of new catalytic systems for the homo- and copolymerization of heteroatom-containing monomers capable of chain configuration regulation

- Inexpensive and straightforward activators to increase the catalyst activity several folds

- Highly active single-site catalysts for the industrial production of polypropylene with competitive cost, specific properties (e.g., elastomeric) and stereoregularity compared to the latest generation of Ziegler-Natta catalysts

- Simple, economical, and well-controlled approaches for the production of in-reactor polymer alloys by mixed catalysts

- New REACH and FDA approved catalytic systems

- Single-site catalytic systems producing polyolefins with enhanced processability

7. Conclusion

Olefins are cost-effective, highly available platforms for the development of functional polyolefins extending the alkenes' value chain. In this chapter, we have discussed how specialty polyolefins are finding their way in niche markets and their potential in shaping future applications. Moreover, the trends and challenges in the production of functional polyolefins are reviewed. Although this class of thermoplastics are synthesized using radical polymerization, at present, the poor control over the microstructural characteristics may boost the insertion polymerization application. Single-site catalysts will be the core in the progression of novel tailor-made functional polyolefins.

Author details

Zahra Balzade, Farhad Sharif* and Seyed Reza Ghaffarian Anbaran
Polymer Engineering and Color Technology Department, Amirkabir University of Technology, Tehran, Iran

*Address all correspondence to: sharif@aut.ac.ir

References

[1] Sauter, D.W., Taoufik, M., Boisson, C., Polyolefins, a Success Story. Polymers. 2017; 9:185. DOI:10.3390/polym9060185.

[2] Hutley, T.J., Ouederni, M., *Polyolefins—The History and Economic Impact*, in *Polyolefin Compounds and Materials: Fundamentals and Industrial Applications*, Al-Ali AlMa'adeed, M., Krupa, I., Editors. 2016, Springer International Publishing: Cham. p. 13-50.

[3] Plastics Market Size, Growth & Analysis [Internet]. 2020. Available from: https://www.reportsanddata.com/report-detail/plastics-market [Accessed: 2021-06-01].

[4] Makaryan, I.A., Sedov, I.V., Analysis of the State and Development Prospects of the Industrial Catalysts Market for Polyolefins Production. Russian Journal of General Chemistry. 2020; 90:1141-1162. DOI:10.1134/S1070363220060304.

[5] Chung TC, M., Polyolefin from Commodity to Specialty. Journal of Material Sciences & Engineering. 2015; 4:1-2. DOI:10.4172/2169-0022.1000e111.

[6] Hoff, R., Handbook of Transition Metal Polymerization Catalysts. Wiley; 2018.

[7] Nowlin, T.E., Business and Technology of the Global Polyethylene Industry: An In-depth Look at the History, Technology, Catalysts, and Modern Commercial Manufacture of Polyethylene and Its Products. Wiley; 2014.

[8] Kaminsky, W., Polyolefins: 50 years after Ziegler and Natta I: Polyethylene and Polypropylene. Springer Berlin Heidelberg; 2013.

[9] Kaminsky, W., Polyolefins: 50 years after Ziegler and Natta II: Polyolefins by Metallocenes and Other Single-Site Catalysts. Springer Berlin Heidelberg; 2013.

[10] Walsh, D.J., Hyatt, M.G., Miller, S.A., Guironnet, D., Recent Trends in Catalytic Polymerizations. ACS Catalysis. 2019; 9:11153-11188. DOI:10.1021/acscatal.9b03226.

[11] Franssen, N.M.G., Reek, J.N.H., de Bruin, B., Synthesis of functional 'polyolefins': state of the art and remaining challenges. Chemical Society Reviews. 2013; 42:5809-5832. DOI:10.1039/C3CS60032G.

[12] Chikkali, S.H., Ziegler–Natta polymerization and the remaining challenges. Resonance. 2017; 22: 1039-1060. DOI:10.1007/s12045-017-0570-2.

[13] Rolando, T.E., Solvent-Free Adhesives. Rapra Technology; 1998.

[14] Long, Y., Pan, Y., Xu, Y., Yan, Y., Modified polybutene-1 film with high dielectric constant based on low doping amount of fluorine-based compound. Materials Today Communications. 2021; 26:102014. DOI:10.1016/j.mtcomm.2021.102014.

[15] Burnett, C.L., Bergfeld, W.F., Belsito, D.V., Hill, R.A., Klaassen, C.D., Liebler, D.C., Marks, J.G., Jr., Shank, R.C., Slaga, T.J., Snyder, P.W., Gill, L.J., Heldreth, B., Safety Assessment of Polyene Group as Used in Cosmetics. Int J Toxicol. 2020; 39:59s-90s. DOI:10.1177/1091581820952385.

[16] Patil, A., Sandewicz, R.W., *Cosmetic Science and Polymer Chemistry: Perfect Together*, in *Polymers for Personal Care and Cosmetics*. 2013, ACS Symposium Series; American Chemical Society: Washington, D.C.. p. 13-37.

[17] Younes, M., Aquilina, G., Castle, L., Engel, K.-H., Fowler, P., Fürst, P.,

Gürtler, R., Gundert-Remy, U., Husøy, T., Mennes, W., Moldeus, P., Shah, R., Waalkens-Berendsen, I., Wölfle, D., Boon, P., Crebelli, R., Di Domenico, A., Filipič, M., Mortensen, A., Van Loveren, H., Woutersen, R., Gergelova, P., Giarola, A., Lodi, F., Frutos Fernandez, M.J., Re-evaluation of name of hydrogenated poly-1-decene (E 907) as food additive. EFSA Journal. 2020; 18:6034. DOI:10.2903/j.efsa.2020.6034.

[18] Safty evaluation of certain food additive and contaminants [Internet]. 2000. Available from: http://www.inchem.org/documents/jecfa/jecmono/v44jec02.htm [Accessed: 2021-06-01].

[19] TPX™ (Polymethylpentene) [Internet]. Available from: https://eu.mitsuichemicals.com/service/mobility/polymers/tpx/index.htm [Accessed: 2021-05-01].

[20] Markova, S., Shalygin, M., Pelzer, M., Gries, T., Teplyakov, V., Application prospects of dense gas separation hollow fibers based on poly(4-methyl-1-pentene). Chemical Papers. 2020; 74:1917-1921. DOI:10.1007/s11696-019-01043-x.

[21] VESTENAMER® the world's most versatile rubber additive [Internet]. Available from: https://www.vestenamer.com/en/rubber-additive [Accessed: 2021-06-05].

[22] Lu, Y., Peng, Z.-l., Liang, H.-m., Zhang, Y.-x., Structure and properties characterization for a novel propylene-ethylene copolymer Vistamaxx. China Elastomerics. 2007; 17:9.

[23] ExxonMobil extends Vistamaxx line with low-viscosity processing aid. Additives for Polymers. 2017; 2017:4. DOI:10.1016/S0306-3747(17)30005-2.

[24] Benefits of Vistamaxx™ performance polymers in color masterbatch formulations, Webinar [Internet]. 2020. Available from: https://www.exxonmobilchemical.com/en/exxonmobil-chemical/webinars[Accessed: 2021-06-01].

[25] Vistamaxx™ performance polymers [Internet]. Available from: https://www.exxonmobilchemical.com/en/products/polymer-modifiers/vistamaxx-performance-polymers [Accessed: 2021-05-10].

[26] Gallez, N.D.V., Vistamaxx™ Propylene-based Elastomers for Application in Hygiene Absorbent Products. China Textile Leader. 2010.

[27] Rudnick, L. R., Lubricant Additives: Chemistry and Applications. 3 ed. CRC Press; 2017.

[28] Lagona, J., Dispersant Viscosity Index Improver-containing Lubricant Compositions And Methods Of Use Thereof, 2018, US 2018/0105764 A1.

[29] Mathur Naresh, C., Duggal, A., Devlin Mark, T., Functionalized Olefin Copolymer Additive Composition, 2009, US 2009/0325831 A1.

[30] Esche Jr Carl, K., West Charles, T., Hybridized Olefin Copolymer Additives, 2006, US 7049273 B2.

[31] Cheremisinoff, N.P., Spotlight on EPDM Elastomers. Polymer-Plastics Technology and Engineering. 1992; 31:713-744. DOI:10.1080/03602559208017776.

[32] Ravishankar, P.S., Treatis on EPDM. Rubber Chemistry and Technology. 2012; 85:327-349. DOI:10.5254/rct.12.87993.

[33] Wu, S., Han, T., Thakur, V., Weeks, R., Li Pi Shan, C., Tuberquia, J., Ultra High Molecular Weight NORDEL™ EPDM for TPV and Low Hardness Molded Goods Enabled by Advanced Molecular Catalyst. IOP Conference Series: Materials Science and Engineering. 2019; 548:012013. DOI:10.1088/1757-899x/548/1/012013.

[34] Bhattacharya, A.B., Raju, A.T., Chatterjee, T., Naskar, K., Development and characterizations of ultra-high molecular weight EPDM/PP based TPV nanocomposites for automotive applications. Polymer Composites. 2020; 41:4950-4962. DOI:10.1002/pc.25765.

[35] Valentini, F., Dorigato, A., Fambri, L., Pegoretti, A., Novel EPDM/paraffin foams for thermal energy storage applications. Rubber Chemistry and Technology. 2021; DOI:10.5254/rct.21.79976.

[36] Valentini, F., Fambri, L., Dorigato, A., Pegoretti, A., Production and Characterization of TES-EPDM Foams With Paraffin for Thermal Management Applications. Frontiers in Materials. 2021; 8: 660656. DOI:10.3389/fmats.2021.660656.

[37] Zhang, R., Kim, E.S., Romero-Diez, S., Wang, Y., Huang, G., Li, A., Yang, Y., Lee, P.C., Cyclic olefin copolymer foam: A promising thermal insulation material. Chemical Engineering Journal. 2021; 409:128251. DOI:10.1016/j.cej.2020.128251

[38] Hlobik, T., Polymer Syringe Considerations for Drug Applications and Administration. ONdrugDelivery. 2020; 90-93.

[39] Konishi, Y., Sawaguchi, T., Kubomura, K., Minami, K., High performance cyclo olefin polymer ZEONEX. Proc. SPIE 5872, Advancements in Polymer Optics Design, Fabrication, and Materials. 2005; 587203. DOI: 10.1117/12.613555.

[40] Kohara, T., Development of new cyclic olefin polymers for optical uses. Macromolecular Symposia. 1996; 101:571-579. DOI:10.1002/masy.19961010163.

[41] Alternative for SARS-CoV-2 Vaccine Primary Package Systems: Daikyo Crystal Zenith® Cyclic Olefin Polymer Vials [Internet]. 2020. Available from: https://www.westpharma.com/ [Accessed: 2021-06-15].

[42] New high-performance thermoplastics for next-generation [Internet]. 2019. Available from: https://www.zeon.co.jp/en/business/enterprise/resin/pdf/200323391.pdf [Accessed: 2021-05-10].

[43] Shieh, P., Zhang, W., Husted, K.E.L., Kristufek, S.L., Xiong, B., Lundberg, D.J., Lem, J., Veysset, D., Sun, Y., Nelson, K.A., Plata, D.L., Johnson, J.A., Cleavable comonomers enable degradable, recyclable thermoset plastics. Nature. 2020; 583:542-547. DOI:10.1038/s41586-020-2495-2.

[44] Lam, H., *Factors Enhancing Adhesion of Color Cosmetic Products to Skin: The Role of Pigments and Fillers*, in *Surface Science and Adhesion in Cosmetics*. 2021. p. 487-541.

[45] Cosmetics Ingredients Selector [Internet]. Available from: https://cosmetics.specialchem.com/ [Accessed: 2021-05-15].

[46] Resins and Film Formers [Internet]. Available from: http://www.koboproductsinc.com/products.aspx?p=Resins&c=Resins%20Oil%20Phase#KOBOGUARD5400-CCT [Accessed: 2021-05-10].

[47] Schellenberg, J., Syndiotactic Polystyrene: Synthesis, Characterization, Processing, and Applications. Wiley; 2009.

[48] Baugh, L.S., Schulz, D.N., Discovery of Syndiotactic Polystyrene: Its Synthesis and Impact. Macromolecules. 2020; 53:3627-3631. DOI:10.1021/acs.macromol.0c00350

[49] XAREC Syndiotactic Polystyrene [Internet]. Available from: https://www.

idemitsu.com /en/business/ipc/ engineering/polystyrene.html [Accessed: 2021-05-05].

[50] Petrukhina, N.N., Tsvetkov, O.N., Maksimov, A.L., Hydrogenated Styrene–Diene Copolymers as Thickening Additives to Lubricating Oils. Russian Journal of Applied Chemistry. 2019; 92:1179-1189. DOI:10.1134/S1070427219090015.

[51] Hyosung Poketone [Internet]. Available from: http://www.poketone.com/ [Accessed: 2021-06-10].

[52] Soomro, S.S., Cozzula, D., Leitner, W., Vogt, H., Müller, T.E., The microstructure and melt properties of CO–ethylene copolymers with remarkably low CO content. Polymer Chemistry. 2014; 5:3831-3837. DOI:10.1039/C3PY01637D.

[53] Morgen, T.O., Baur, M., Göttker-Schnetmann, I., Mecking, S., Photodegradable branched polyethylenes from carbon monoxide copolymerization under benign conditions. Nature Communications. 2020; 11:3693. DOI:10.1038/s41467-020-17542-5.

[54] Dai, Y., Luo, J., Liu, T., Jia, L., Dual-site catalysis for sustainable polymers to replace current commodity polymers – carbonylative copolymerization of ethylene, ethylene oxide, and tetrahydrofuran. Chemical Communications. 2020; 56:15341-15344. DOI:10.1039/D0CC05657J.

[55] Exfola™ Surface Modifier [Internet]. Available from: https://us.mitsuichemicals.com/service/product/exfola.htm [Accessed: 2021-05-12].

[56] EMAC® and EBAC® Acrylate Resins [Internet]. Available from: https://www.westlake.com/polyethylene-products/emac%C2%AE-and-ebac%C2%AE-acrylate-resins [Accessed: 2021-05-10].

[57] LOTRYL® resin, Ethylene-Acrylate copolymers [Internet]. Available from: https://www.lotryl.com/en/ [Accessed: 2021-05-16].

[58] Zhong, S., Tan, Y., Zhong, L., Gao, J., Liao, H., Jiang, L., Gao, H., Wu, Q., Precision Synthesis of Ethylene and Polar Monomer Copolymers by Palladium-Catalyzed Living Coordination Copolymerization. Macromolecules. 2017; 50:5661-5669. DOI:10.1021/acs.macromol.7b01132.

[59] Xiong, S., Shoshani, M.M., Zhang, X., Spinney, H.A., Nett, A.J., Henderson, B.S., Miller, T.F., Agapie, T., Efficient Copolymerization of Acrylate and Ethylene with Neutral P, O-Chelated Nickel Catalysts: Mechanistic Investigations of Monomer Insertion and Chelate Formation. Journal of the American Chemical Society. 2021; 143:6516-6527. DOI:10.1021/jacs.1c00566.

[60] NUCREL™ AE Acid Copolymer [Internet]. Available from: https://www.dow.com/en-us/pdp.nucrel-ae-acid-copolymer.1891973z.html [Accessed: 2021-06-11].

[61] Shahinpoor, M., Fundamentals of Smart Materials. Royal Society of Chemistry; 2020.

[62] Francesconi, A., Giacomuzzo, C., Grande, A.M., Mudric, T., Zaccariotto, M., Etemadi, E., Di Landro, L.,Galvanetto, U., Comparison of self-healing ionomer to aluminium-alloy bumpers for protecting spacecraft equipment from space debris impacts. Advances in Space Research. 2013; 51:930-940. DOI:10.1016/j.asr.2012.10.013.

[63] Fallahi, A., Bahramzadeh, Y., Tabatabaie, S.E., Shahinpoor, M., A novel multifunctional soft robotic transducer made with poly (ethylene-co-methacrylic acid) ionomer metal nanocomposite. International Journal of

Intelligent Robotics and Applications. 2017; 1:143-156. DOI:10.1007/s41315-017-0013-y.

[64] Azarian, M.H., Wootthikanokkhan, J., Gelatin-based solid electrolytes for chromogenic windows applications: a review. Ionics. 2020; 26:5841-5851. DOI:10.1007/s11581-020-03774-4.

[65] Evatane® [Internet]. Available from: https://www.evatane.com/en/evatane/ [Accessed: 2021-05-10].

[66] EVA [Internet]. Available from: https://www.lyondellbasell.com/en/products-technology/polymers/resin-type/eva/ [Accessed: 2021-05-05].

[67] LEVAPREN® [Internet]. Available from: https://www.arlanxeo.com/en/brands/levapren [Accessed: 2021-06-10].

[68] Sumitomo Chemical Develops New Encapsulant Material for Solar Panels - Reducing PID with high transparency unaffected [Internet]. 2014. Available from: https://www.sumitomo-chem.co.jp/english/news/detail/20141003.html [Accessed: 2021-06-10].

[69] Orevac® T [Internet]. Available from: https://www.ambrapolymers.com/en/products/orevac_t.html [Accessed: 2021-06-12].

[70] Balzadeh, Z., Arabi, H., Novel phenolic antioxidant-functionalized dendritic polyethylene: Synthesis by tailor-made nickel(II) α-diimine-catalyzed copolymerization and its characteristics as non-releasing additive. Reactive and Functional Polymers. 2017; 111:68-78. DOI:10.1016/j.reactfunctpolym.2016.11.005.

[71] Balzadeh, Z., Arabi, H., Insights into the chemical composition and thermo-oxidative stability of novel polyethylene copolymers containing ancillary phenolic antioxidant groups as non-migrating polyolefin stabilizer. Polymer Degradation and Stability. 2017; 142:139-149. DOI:10.1016/j.polymdegradstab. 2017.05.034.

[72] Luckham, S.L.J., Nozaki, K., Toward the Copolymerization of Propylene with Polar Comonomers. Accounts of Chemical Research. 2021; 54:344-355. DOI:10.1021/acs.accounts.0c00628.

[73] Kaminsky, W., Trends in Polyolefin Chemistry. Macromolecular Chemistry and Physics. 2008; 209:459-466. DOI:10.1002/macp.200700575.

[74] Hassanian-Moghaddam, D., Maddah, Y., Ahmadjo, S., Mortazavi, S.M.M., Sharif, F., Ahmadi, M., Mechanistic study on the metallocene-based tandem catalytic coordinative chain transfer polymerization for the synthesis of highly branched polyolefins. European Polymer Journal. 2021; 152:110454. DOI:10.1016/j.eurpolymj.2021.110454.

[75] Maddah, Y., Ahmadjo, S., Mortazavi, S.M.M., Sharif, F., Hassanian- Moghaddam, D., Ahmadi, M., Control over Branching Topology by Introducing a Dual Catalytic System in Coordinative Chain Transfer Polymerization of Olefins. Macromolecules. 2020; 53:4312- 4322. DOI:10.1021/acs.macromol. 0c00358.

[76] Kay, C.J., Goring, P.D., Burnett, C.A., Hornby, B., Lewtas, K., Morris, S., Morton, C., McNally, T., Theaker, G.W., Waterson, C., Wright, P.M., Scott, P., Polyolefin–Polar Block Copolymers from Versatile New Macromonomers. Journal of the American Chemical Society. 2018; 140:13921-13934. DOI:10.1021/jacs.8b09039.

[77] Ntetsikas, K., Zapsas, G., Bilalis, P., Gnanou, Y., Feng, X., Thomas, E.L., Hadjichristidis, N., Complex Star Architectures of Well-Defined Polyethylene-Based Co/Terpolymers. Macromolecules. 2020; 53:4355-4365. DOI:10.1021/acs.macromol. 0c00668.

Section 5

Industrial Aspects

Chapter 6

A Look at the Industrial Production of Olefins Based on Naphtha Feed: A Process Study of a Petrochemical Unit

Reza Davarnejad, Jamal Azizi and Shaghayegh Bahari

Abstract

Olefins (ethylene, propylene and butadiene) as raw materials play an important role in a lot of chemical and polymer products. In industrial scale, there are several techniques from crude oil, natural gas, coal and methanol for the olefins production. Each of these has some advantages. The petrochemicals with liquid feed can simultaneously produce all of the olefins. Shazand Petrochemical Co. (as the first olefins production unit in Iran) produces all of the olefins using naphtha (light and heavy) feed. In this chapter, the production process of olefins based on naphtha will be studied from the beginning to the end (involving pyrolysis, compression, chilling and fractionation processes).

Keywords: Naphtha, Olefin unit, Petrochemical industry, Separation process

1. Introduction

1.1 Olefins

Olefins are unsaturated compounds with a formula of CnH_2n. The names of these compounds end with –ene, such as ethene (ethylene), propene (propylene), butenes and butadiene. Olefin species are not commonly found in the crude oil. Ethylene and propylene are known as important sources of various industrial chemicals and plastics products. Butadiene is widely used in synthetic rubber production. These as light olefins are industrially produced by pyrolysis and fluid catalytic cracking of the vacuum distillates. Another potential technique for the light olefins production is direct conversion of syngas [1–4].

1.2 Shazand Petrochemical Co.

Shazand Petrochemical Complex produces different range of petrochemical products such as plastics, synthetic rubber and the other chemicals from naphtha (as feedstock). Its total saleable annual production capacity presently is around 620000 tons achieved in eighteen plants to serve domestic and global markets with the standard grades [5]. **Figure 1** shows flow diagram of the Shazand Petrochemical Co. products.

IntechOpen

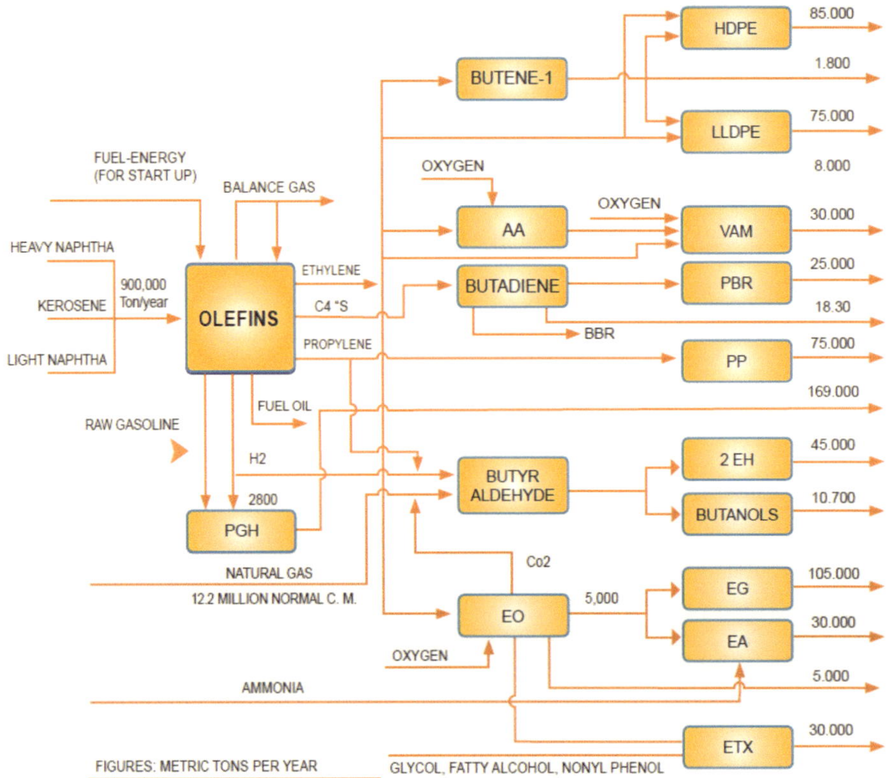

Figure 1.
Flow diagram of the products of Shazand Petrochemical complex [5].

2. Olefin plant

Olefin plants are large, complex units at the heart of petrochemical complexes. Olefin production is a petrochemical process in which saturated hydrocarbons are broken down into smaller, often unsaturated ones. It is the principal industrial method for producing the lighter alkenes commonly known as olefins, including ethene or ethylene, propene or propylene and butadiene. The heavy hydrocarbon compounds are usually cracked at temperatures between 800°C and 860°C. The resulting gas is separated into valuable products suitable for downstream processes. The separation train of a cracker starts with the Hot Section and then the cracked gas is quenched and cooled where heavy fractions are condensed and separated. Later the process steam is condensed and removed from the cracked gas before it enters the compression and chilling sections (Cold Section). The compressed, cracked gas is then separated into specified products. Some fractions (e.g. ethane, propane and butane) are recycled to the furnaces in order to improve overall yields. Today's modern ethylene plants are complex networks of more than 300 individual units, including thermal cracking, cracked gas compression and physical scrubbing, fractionation, adsorptive drying, catalytic hydrogenation and others, operating in the temperature range of 1100°C to −170°C [6–9].

2.1 Hot section

The goal of this section is to take the feed from the sources and pyrolysis it in the furnaces and convert it to organic radicals and hydrocarbons such as hydrogen, propane, ethane, ethylene, gasoline, butane, propylene and other heavy compounds which gasoline and fuel oil are the first olefin products. Hot Section consists of six subunits:

- Feed Preheating.
- Cracking Furnaces.
- Primary Fractionation and Fuel Stripping.
- Quench Water/Gasoline System.
- Dilution Steam Generation.
- Low Pressure Steam (LPS) Generation.

2.1.1 Industrial steam cracking

Steam cracking furnaces are process units, devoted to producing ethylene and propylene from a stream of hydrocarbons and steam. Thermal cracking, pyrolysis, with water vapor is a good way to convert inactive paraffinic hydrocarbons to olefin active compounds, which are the main feed for the petrochemical industry. The raw materials used in thermal cracking processes include ethane, propane and butane, liquid naphtha and diesel. The liquid feed of the olefin unit includes light naphtha to heavy diesel. The most important part of an olefin unit is where pyrolysis reactions occur. This part is the reactor inside a heat furnace. Numerous factors such as the amount and type of feed, operating temperature, pressure, the ratio of diluent vapor to the inlet feed, the amount of coke settling, the material and shape of the reactor are effective in the efficiency of products in thermal reactors. In steam cracking, naphtha is diluted with steam and briefly heated in a furnace in the absence of oxygen. Typically, the reaction temperature is very high, around 850°C. The reaction occurs rapidly as the residence time is around some milliseconds and thus the flow rate approaches the speed of sound. When the cracking temperature has been reached, the gas then is quickly quenched in a transfer line heat exchanger or inside a quenching header using quench oil to stop the reaction [10–15].

Pyrolysis as a complex phenomenon has two main parts:

- Dehydrogenation.
- Breaking the C-C bond and converting it to free radicals.

At high temperatures, hydrocarbons become unstable and decompose to hydrogen, methane, C=C and aromatic compounds. The higher the temperature is, the more olefin and aromatic compounds are formed high stability. Therefore, not only light olefin compounds such as ethylene and propylene, but also heavier compounds such as aromatics and tar will be produced.

Research has shown that the cracking reaction is of the first order. The amount of cracking does not depend on pressure. Additionally the activation energy for these reactions is significantly lower than the energy required breaking the C-C bond. Thus, the conversion can be considered as a single molecular reaction and follow the mechanism of first order reactions. According to Rice and Herzfeld, the reaction mechanism is based on the formation of radicals and is a series of chain reactions that result in the production of olefins from hydrocarbons. The mechanism of chain reactions also explains the low activation energy of these reactions. To be more precise, we consider the following reactions:

a. Break the C-C bond in ethane and turn it into two radicals:

$$C_2H_6 \rightarrow \dot{CH}_3 + \dot{CH}_3 \quad (1)$$

b. Joining and combining a radical with a feed molecule:

$$\dot{CH}_3 + C_2H_6 \rightarrow CH_4 + C_2\dot{H}_5 \quad (2)$$

c. Emission stage: Here the new radical is converted to an olefin and hydrogen radical. The hydrogen radical later collides with another feed molecule to form a new radical and a hydrogen molecule:

$$\dot{C}_2H_5 \rightarrow C_2H_6 + \dot{H} \quad (3)$$

$$\dot{H} + C_2H_6 \rightarrow H_2 + C_2\dot{H}_5 \quad (4)$$

d. Final stage: Chain reactions end in the following four ways:

$$\dot{H} + \dot{CH}_3 \rightarrow H_2 + CH_4 \quad (5)$$

$$\dot{C}_2H_5 + C_2\dot{H}_5 \rightarrow C_4H_{10} \quad (6)$$

$$\dot{H} + \dot{H} \rightarrow H_2 \quad (7)$$

$$\dot{H} + C_2\dot{H} \rightarrow C_2H_2 \quad (8)$$

The final result of chain reactions is that of steps 3 and 4.

$$C_2H_6 \rightarrow C_2H_4 + H_2 \quad (9)$$

In the final stage, methane and heavier compounds such as butane are formed. Speed and rate of reaction transformations:

According to the Arrhenius equation, the reaction rate constant is $k = A.\exp.^{-E/RT}$ where A is Arrhenius constant, E is activation energy, R is universal Gas constant and T is temperature.

Since the reaction is a first order, the rate of conversion is obtained from the equation $kt = \ln 1/1\text{-}x$.

Where t is the time and x is the amount of conversion of raw materials to crack gas product. This equation is only available at some temperatures because k depends on the temperature. The pressure must be low because at high pressures the reaction follows the second order. Therefore, dilution steam is used to keep the partial pressures low. Conversion of feed to a high quality ethylene product depends on the temperature rising. Temperature should linearly be increased and within a certain range. That is because at high temperatures side effects may occur and these reactions will be of the second-order type, it is likely that ethylene will be converted to methane, which reduces efficiency [16–22].

2.1.2 Cracking furnaces

In the olefin unit, the furnace plays the main role. In general, its job is thermal cracking by which heavy hydrocarbons are broken down and hydrocarbons such as propane, ethane, ethylene, propylene, hydrogen, C4 and C4+ are produced. **Figure 2** shows how feed compound is cracked in the furnace.

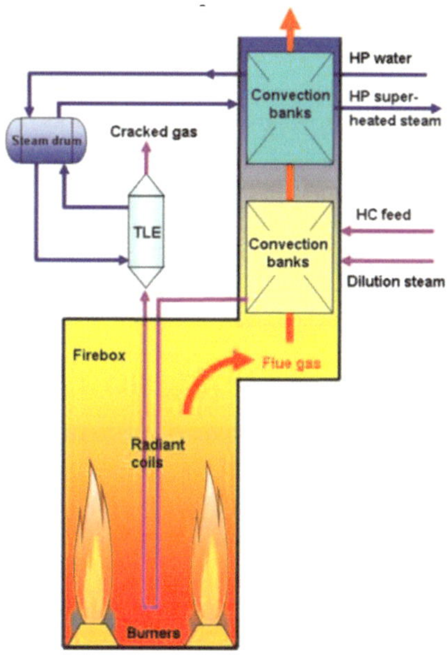

Figure 2.
Schematic of the furnace [23].

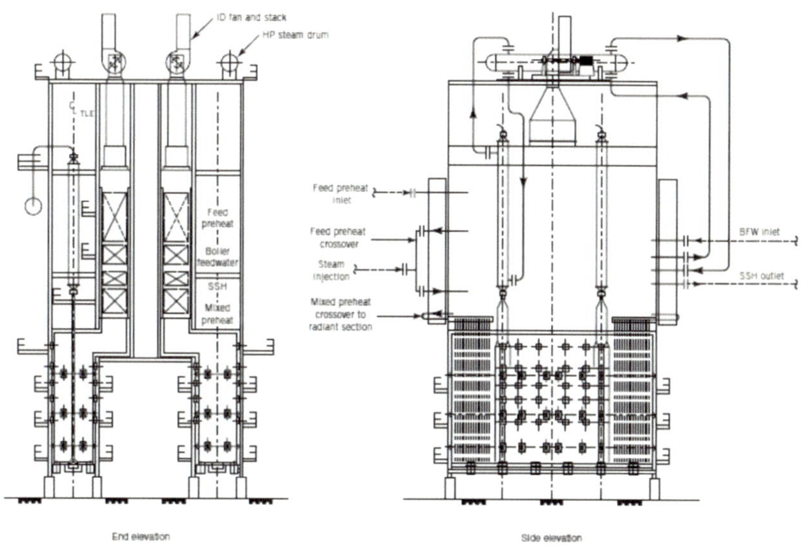

Figure 3.
Typical heater configuration (BFW = boiler feed water; SSH = super high pressure steam; HP = high pressure; and ID = induced draft [24]).

2.1.3 Structure of furnaces

In general, furnaces consist of two parts: Radiation and Convection section. There are 10 furnaces in Shazand Petrochemical Olefin Unit, where there are usually 9 furnaces in service and one in de-cocking mode, which will be replaced by another furnace after de-coking. **Figure 3** depicts a typical configuration of an olefin heater.

2.1.3.1 Radiation section (fire box)

This part occupies a larger volume of furnaces and includes 8 coils which are installed in the form of vertical pipes and have 4 coils on both sides. All coils are collected at the end by a collector outlet and the compounds are discharged through a transmission line at the end of the Fire Box. 16 streams are sent to this section through the outlet pipes of the convection section. 108 burners are located on either side of the Radiation Section, which premix fuel and air before reaching the nozzle head. The fuel used for these burners is methane.

2.1.3.2 Convection section

The heat generated in the radiation section is used in the convection section to pre-heat the input feed. The temperature profile in the Radiation Section increases from the bottom to the top, but in the convection section it decreases from the bottom to the top. Therefore, the highest temperature is at the top of the Fire Box. Hence, in general, this type of furnace can be named as reactor which the heated feed enters from above and breaks at the bottom.

2.1.4 Cracking furnace performance

After pre-heating and passing through the emergency valve, the naphtha feed enters the convection section of the furnace. In furnaces, the temperature gradually rises in several stages by coils. The feed temperature rises from 75–105°C and is converted from a liquid to a vapor and mixed with the pre-heated recycled propane and ethane. The diluent steam at 363°C is added to the feed mixture to prevent the hydrocarbon vapor pressure increasing while the mixture's temperature reaches 172°C. Now, the mixture enters the furnace and reaches 603°C in the high temperature coil. The feed flows from this part to the radiation part inside 8 coils which are arranged in vertical tubes and cracking is performed. It finally exits the radiation section with a temperature of 863°C and because side reactions may occur, it quickly enters a heat exchanger called TLE or TLX by which it is cooled with water down to 325°C. The pressure of the mixture is 7.3 bar.

TLE heat exchangers are in the form of shell and tubes which water passes through the shell to cool the cracked gas inside the tubes. Because this temperature is still high, another heat exchanger (called quench fitting) is used to reduce the temperature, which quench oil is mixed with the feed stream and the feed temperature is reduced to 165°C, when the mixture is ready for the next stage (called primary fractionation).

Due to the continuous operation of the furnaces, they must be decontaminated periodically. Therefore, the furnace must be in de-cocking mode approximately every 83 days [23, 25, 26].

2.2 Cold section

The purpose of this section is to separate the following materials from the cracked gas that comes from the Hot Section.

1. Separation of hydrogen and methane

2. Ethylene separation

3. Separation of propylene

4. Separation of C4 compounds

5. Separation of gasoline and heavier compounds

Cold Section consists of compressors and low temperature fractionation sections.

2.2.1 Compressors

In the Cold Section, there are three compressors, two of them are called the Propylene (C-501) and the Ethylene (C-502) compressors, which have a closed cycle and operate according to the Rankin cycle and their task is to cool down the cracked gas to −36°C and − 93°C, respectively. The third compressor (C-201), with a higher capacity than the previous two is utilized to compress the cracked gas from the furnaces where contains the compounds of hydrogen, carbon monoxide, carbon dioxide, acetylene, ethylene, ethane, propadiene, propylene, propane, vinyl acetylene, butadiene, butane, butene, C4+ and water.

First the cracked gas enters a vessel (V-201) where is actually the suction drum of the first stage of a five-stages compressor at 12°C and a pressure of 0.42 bar. Gases escape from the top of the V-201 and liquids (oily water) from the bottom.

Exhaust gases from this tank enter the first stage of the C-201 compressor. The output of this section with a pressure of 3 bar and a temperature of 85°C enters the cooling exchanger (E-201) losing heat to cooling water stream and then exits the exchanger at 35°C and enters to second stage suction drum (V-202). This vessel has a temperature of 29°C and a pressure of 3 bar and at its bottom contains oily water which exits and joins the previous oily water flow from the V-201. In this vessel, pyrolysis gasoline is separated and sent to the tower (T-201). Light gas from the top of the tower is returned to V-201 and semi-light gasoline from the bottom of the tower is sent to a storage tank with pressure and temperature of 0.5 bar and 90°C respectively. There are three more stags in the compressor compressing the cracked gas to 10, 20 and 37 bar, respectively. Other vessels (V-203, 204, 220, 206 and 207) also separate liquids from gas as the compressor's suction drums. The exhausted gas from the third stage must be sweetened and its sulfur compounds and carbon dioxide must be removed by a tower (T-202), where the absorption operation is performed using a caustic solution. In order to prevent the condensation of hydrocarbons and the formation of polymers the output of the V-204, first enters the heat exchanger (E-215) where it reaches a temperature of 45°C and then it enters the tower (T-202).

2.2.1.1 Caustic wash tower (T-202)

The tower consists of three parts: the bottom part of the tower, which performs about 70% of sweetening while the middle part performs about 30% of sweetening operations. At the top of the tower, boiler feed water (B.F.W) is supplied for top-down washing of cracked gas. The feed enters the tower from the first tray of bottom and the water from the top and the caustic solution is continuously circulated inside the tower for more efficiency. The consumed caustic comes out from the bottom of the tower for treatment. Sometimes it is necessary to take the lower part of the tower out of service and wash it to remove the annoying materials and formed polymers. In these cases, feed will enter from the middle part.

Finally, the exhausted gas from the compressor is cooled by a propylene cycle and enters V-207, where the cracked gas is converted into 2 phases of gas and liquid

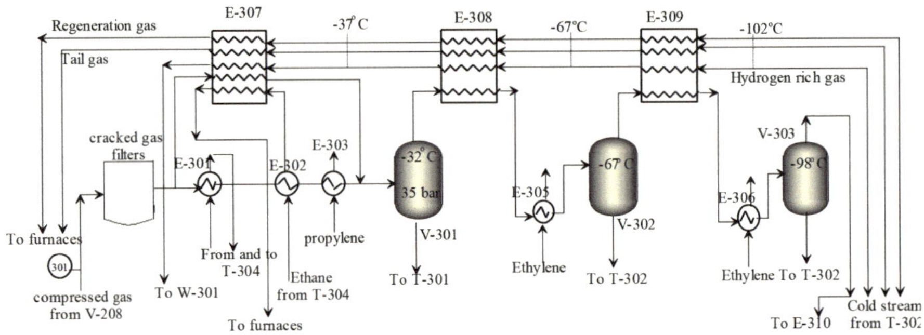

Figure 4.
Chilling unit entrance.

and each phase enters the dryers of the cold section unit separately for drying operations.

2.2.2 Low temperature fractionation

As illustrated in **Figure 4**, this section consists of four plate fin-type heat exchangers (E-307, 308, 309 & 310) which are known as demethanizer feed exchangers cool the cracked gas by separating flows of tail gas, regeneration gas and hydrogen. In fact, the main task of this part is to cool the gas for the feed of demethanizers. The cold section inlet gas first passes through the strainers and reaches −93°C using 5 series kettle-type heat exchangers and also with the help of a cold box. Using flashing in 3 separate vessels (V-301, 302 & 303), feeds for demethanizers (T-301, 302) as well as pressure swing adsorption (PSA) unit are prepared.

2.2.2.1 Hydrogen purification

The purpose of this unit is to provide pure hydrogen for hydrogenation of acetylene, methyl acetylene and propadine in reactors of olefin unit and hydrogenation of pyrolysis gasoline in PGH unit as well as for heavy density polyethylene (HDPE), linear low density polyethylene (LLDPE) and polypropylene (PP) units. 99.99% purity is essential for hydrogen, which is done by PSA. In this unit, the pressure swing adsorption is used to purify hydrogen.

2.2.2.2 Demethanizer and methane separation (T-301 and T302)

Figure 5 shows demethanizer package of olefin plant in Arak petrochemical complex. In demethanizer, methane is removed by fractionation as an over head product. The bottom product, consisting of ethane and heavier compounds, is the feed to the deethanizer. The first demethanizer (T-301) has 20 trays at which input feed enters from tray 16, is supplied from the bottom of the V-301. The bottom temperature of the tower is 17°C, which is supplied by kettle-type re-boiler and quenched water. The top flow cooled by a propylene cycle to −35°C, enters the second demethanizer.

The vapor at the top of T-301 is the main feed for T-302, injected at tray 28, which has 50 trays. The feed of this tower also includes the second, third and fourth chilling stages coming from the bottom of V-302, V-303, V-304 and injected at tray 46. For each of these feeds, it is possible to change the entry point up to two trays higher. The boiling heat is supplied by the condensed propylene (3°C) in the re-

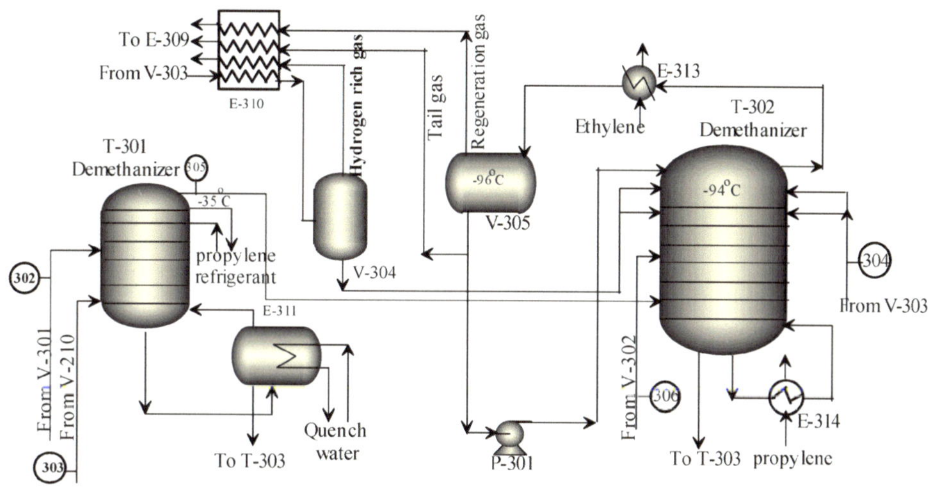

Figure 5.
Demthanizer towers number 1 and 2 (T-301 and T-302).

boiler (E-314), where the flow rate of the propylene is set by the temperature of tray 17 and the bottom product of the tower is sent to the deethanizer (T-303). The impure vapor at the top is cooled by a partial ethylene condenser (−101°C). The pure gas product is sent from V-305 to plate-type exchangers to reach a temperature of 10°C and then it goes to the regeneration system. Part of V-305 liquid is sent to the top of the tower by the pump (P-301) as a reflux.

2.2.2.3 Deethanizer (T-303)

As shown in **Figure 6**, in the deethanizer system of olefin plant, ethane is removed as an overhead product. Propane and heavier components leave the deethanizer as bottom products and are fed to the depropanizer. The bottom products of the first demethanizer (T-301) and the second demethanizer (T-302) enter T-303 separately. Re-boiler heat is generated by low pressure steam (LPS) and the temperature is controlled by the twentieth tray. Exhaust gases are transferred from the top of the tower to the acetylene hydrogenation section. The stream of hydrogenated acetylene returns to the partial condenser (E-315) which is cooled by

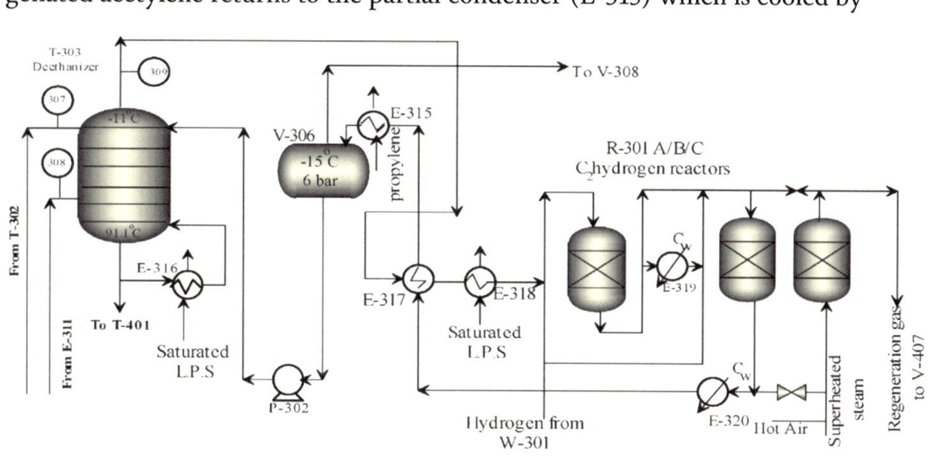

Figure 6.
Deethanizer tower (T-303) and acetylene reactors.

propylene to −20°C and goes to V-306 for separation. The liquid of vessel *i* pumped to T-303 by the P-302 as reflux and the pure gas goes to the C2 splitter tower (T-304).

2.2.2.4 Acetylene hydrogenation

Top products of T-303 first enter E-317 and temperature reaches 45°C and for hydrogenation they enter the first stage of the reactors (R-301 A/B /C) and mix with hydrogen coming from PSA Unit (w-301), the amount of hydrogen is automatically controlled by the feed flow rate. The hydrogenation process temperature on the catalyst bed must be increased (45°C -70°C). To ensure the conversion of all available acetylene, the second reactor is used in series where the output flow of the first reactor is mixed with hydrogen again and enters the second one. To ensure the continuity of operations, three reactors have been designed and two of them are in service and one is in standby mode.

2.2.2.5 Separation of ethylene from ethane (T-304, C2 splitter)

Figure 7 illustrates the loop by which ethylene is separated from ethane as the stream passes through the splitter. Top flow of T-303 enters gas dryer (V-308) and produced water in the reactors is removed. The feed enters T-304, with total 130 trays, at tray 34 or 38. The top 10 trays of this tower are called pasteurization section to separate lighter compounds from ethylene product. Two re-boilers have been provided for this system. Pressure inside the tower varies from about 21.2 to 22.4 bar and the temperature from −27 to −3.4°C. The main re-boiler of the tower is located at the bottom of the tower and uses condensed propylene (at 3°C) for heating. The second re-boiler (E-301) is an internal re-boiler located at tray 39. The condenser of this tower (E-321) works with propylene vapor at −35°C. Top products enter V-309, which is known as Splitter Reflux Accumulator. Off gas is returned to V-206 as a splitter recycle. The liquid of V-309 is also returned to the tower by the pump (P-303) as Reflux. The ethylene product obtains from tray 120 flows to ethylene product surge drum (V-310). The ethylene collected in V-310 has 21.3 bar pressure and a − 27°C temperature. Ethylene product is normally sent to battery limits and after heating. It turns into a gaseous state with a pressure of 21 bar. The excess ethylene is sent to the ethylene tank in liquid form for storage.

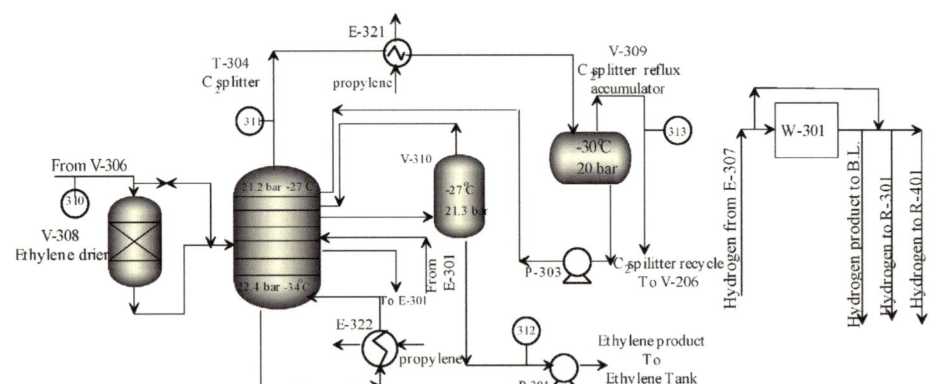

Figure 7.
The configuration of splitter for separation of ethylene from ethane.

The temperature of liquid ethylene in tank is about −128°C and has an atmospheric pressure.

2.2.2.6 Depropanizer (T-401)

As represented in **Figure 8** a depropanizer is a distillation column that is used to separate propane from a mixture containing butane and other heavy components. In the depropanizer, propane is removed as an overhead product and butane and heavier compounds are fed to a debutanizer. This section is known as medium temperature fractionation. The bottom product of T-303 (deethanizer) is mixed with some recycled flow and feed to T-401 at tray 26. The condenser (E-401) works with propylene cycle at 12°C. There are two re-boilers (E-402) that work with low pressure steam (LPS) and change temperature from 42–90°C. The top product of T-401 is collected inside V-401 after condensation (E-401). Part of the liquid is pumped by P-401 to the tower as reflux and another part is directed to the hydrogenation section in R-401 reactors.

2.2.2.7 Debutanizer (T-402)

As shown in **Figure 9**, the debotanizer tower contains 40 trays. The feed enters on tray 21 or 25 directly from the bottom of the depropanizer tower. The vapors of the top of the tower are condensed by the condenser (E-403) and then collected in the accumulator (V-402). By the P-403 one flow is returned to the tower as a reflux and another flow is sent to the battery limits as C4 product. The bottom product of T-402 is light gasoline which is mixed with heavy gasoline, cooled together in E-405 and finally sent for storage.

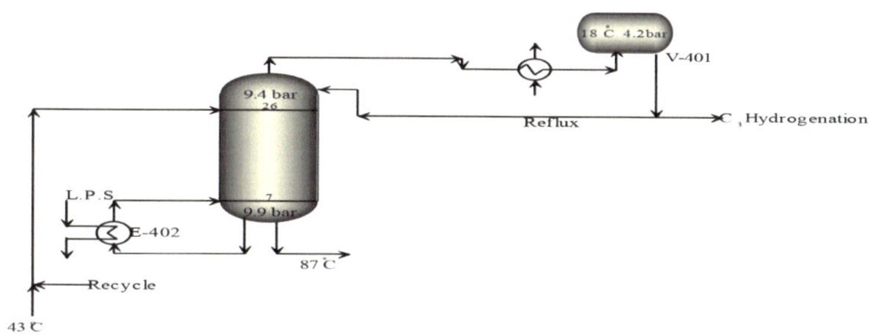

Figure 8.
Depropanizer tower (T-401).

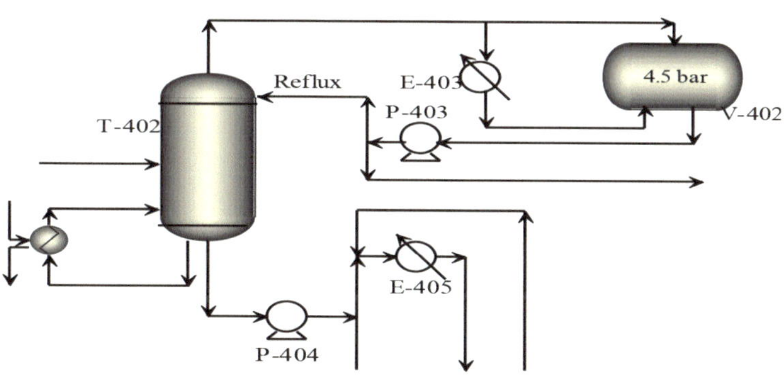

Figure 9.
Debutanizer tower (T-402).

2.2.2.8 Hydrogenation of C_3 cut

The top product of the depropanizer is mixed with recycles and fed to the hydrogenation section in order to convert methyl acetylene and propadiene. The feed is mixed with hydrogen and enters the first reactor. **Figure 10** shows the reactor in which hydrogen is compounded with the feed stream as well as C_3 stripper. For hydrogenation, three reactors have been installed, two in service and one in standby mode. The output of the reactors enters T-403 (**Figure 11**) to separate the light components where the bottom product of this tower enters the C_3 splitter (T-404), with 200 trays, at tray 55 or 68. The tower condenser (E-410) is cooled by cooling water. The produced propane is directed to E-412 for cooling and then to battery limits. A by-product (propane) is taken from tray 4 or 8 and

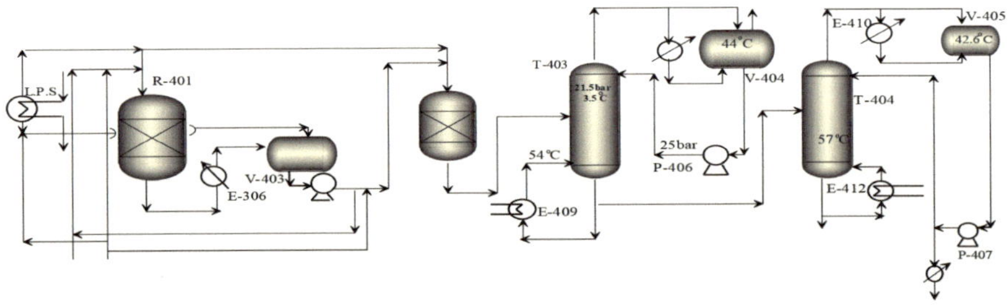

Figure 10.
C_3 reactor and stripper to add hydrogen to the feed stream.

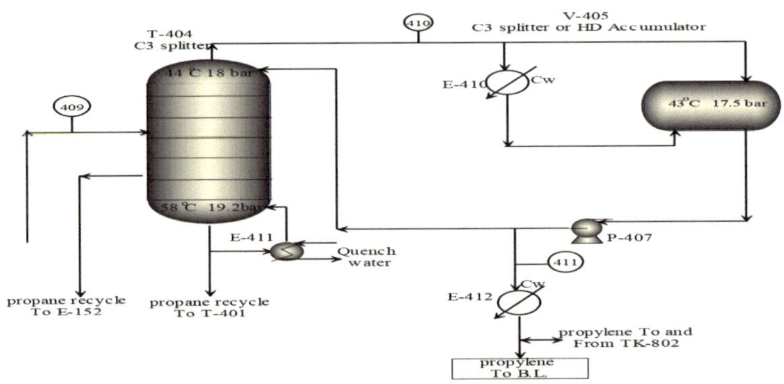

Figure 11.
C3 splitter (T-404).

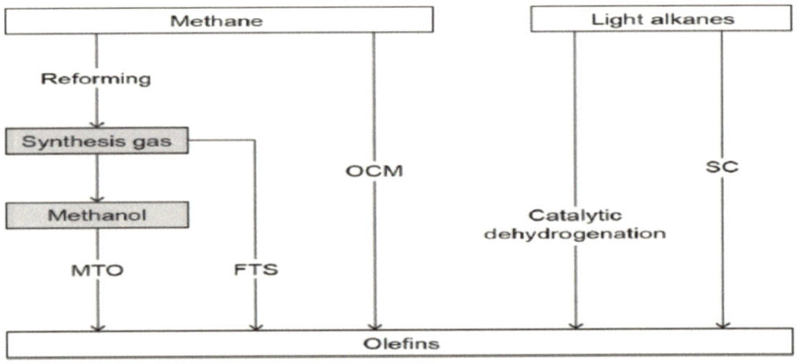

Figure 12.
Different technologies of interest for the production of light olefins from methane and light alkanes [27].

recycled to the furnaces. The tower re-boiler (E-412) works with quench water. The bottom product which contains C_4 and heavier compounds is recycled to the depropanizer (T-401). **Figure 12** illustrates C_3 splitter configuration in olefin plant.

3. New trends in olefin production

In this section, some of the most promising alternatives are compared with the conventional steam cracking process. **Figure 12** depicts novel technologies for the production of light olefins from methane and light alkanes. These technologies emerge especially from the abundance of cheap propane, ethane, and methane from shale gas and stranded gas. Continuing search for alternative and preferably also more sustainable processes and feeds will eventually be required in order to fulfill the future demand for commodity chemicals. The following technologies are of interest: the catalytic dehydrogenation of light alkanes, the oxidative coupling of methane (OCM), and syngas-based routes such as the Fischer-Tropsch synthesis (FTS) and methanol synthesis followed by methanol to olefins (MTO) [27–31].

Biomass is also considered a promising alternative feed that can be converted into the valuable olefins, among other chemicals and fuels (**Figure 13**). Through processes such as fermentation, gasification, cracking and deoxygenation, biomass derivatives can be effectively converted into C_2–C_4 olefins. In this respect, biomass and waste streams are believed to be important for the production of chemicals in the future. In recent years, bio-ethanol has been extensively studied as an alternative feed for C_2H_4 production (**Figure 14**). Other bio-derived compounds such as

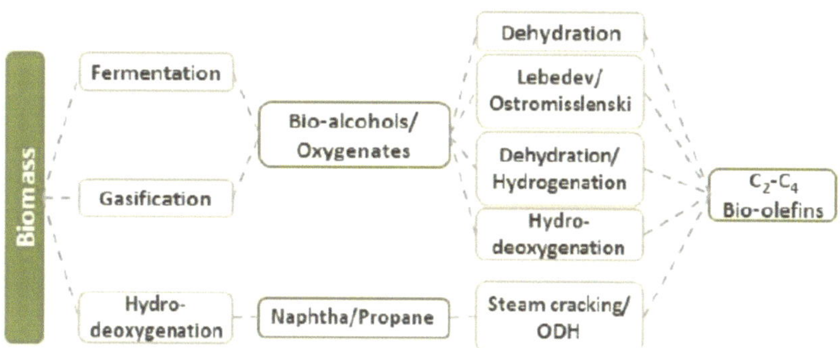

Figure 13.
Biomass to olefins primary routes [32].

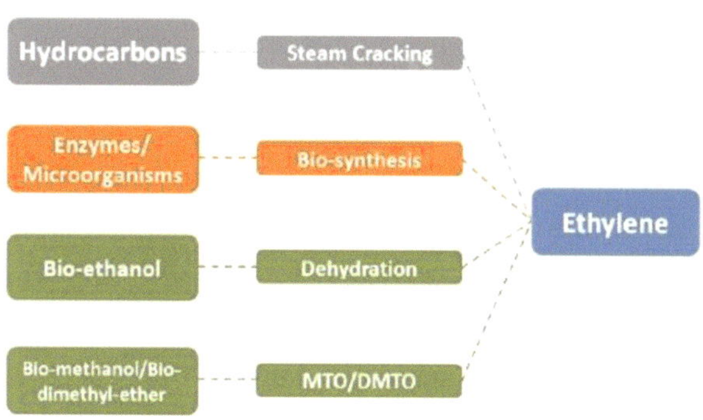

Figure 14.
Schematic chart of C_2H_4 production methods [32].

methanol and dimethyl-ether, can be also be used as a feed for C_2H_4, via methanol to olefins (MTO) and dimethyl-ether to olefins (DMTO) processes. Bio-ethylene can also be produced via bio-synthesis from various enzymes or micro-organisms [32–36] (**Figure 14**). Similarly for production of C_3H_6 and C_4H_6 bio materials have been proven promising using different processes rather than steam cracking.

4. Conclusions

Liquid feed petrochemicals have a high product variety and can produce basic olefins simultaneously. Nevertheless, production costs are higher in liquid feed petrochemicals than in gas based petrochemicals. Because oil is a non-renewable resource, it must be replaced with new technologies in the near future. Methane and ethanol as renewable resources are an appropriate alternative that can be converted into functional olefins using new environmentally friendly technologies.

Author details

Reza Davarnejad*, Jamal Azizi and Shaghayegh Bahari
Department of Chemical Engineering, Faculty of Engineering, Arak University, Arak, Iran

*Address all correspondence to: r-davarnejad@araku.ac.ir

References

[1] https://www.leamericas.com/en/technologies/petrochemical/ethylene-plants/index.html

[2] Rozier, D.E. Thermal DeNOx Optimization: A Case Study in Fluid Catalytic Cracking. Dissertation. University of Mississippi, 2017.

[3] Gonzalez, G.L. Hydrocarbons for Chemical and Specialty Uses, Chapter 11, 2019.

[4] Vogt, E.T.C. Weckhuysen, B.M. Fluid catalytic cracking: recent developments on the grand old lady of zeolite catalysis. Chemical Society Reviews 20, 2015, 7342-7370.

[5] https://www.arpc.ir/Default.aspx?tabid=330&language=en-US

[6] Gary, J.H. et al. Petroleum refining: technology and economics, CRC Press, 2007.

[7] He, W. Benson, R. Applied Plastics Engineering Handbook. (2011) 159-175.

[8] Sanfilippo, D. Miracca, I. and Di Girolamo, M. Engineering Alkanes to Olefins and Higher Value Chemicals, Sustainable Strategies for the Upgrading of Natural Gas: Fundamentals, Challenges, and Opportunities. Springer, Dordrecht, 2005, 217-247.

[9] Abdulrahman A.R. Design and optimization of ethylene production utilizing new technologies. University of Colorado at Boulder, 2001.

[10] Sami, M. Hatch, L.F. Chemistry of Petrochemical Processes, Elsevier, 2001.

[11] Avelino, C. et al. Crude oil to chemicals: light olefins from crude oil, Catalysis Science & Technology 1, 2017, 12-46.

[12] Barazandeh, K. et al. Investigation of coil outlet temperature effect on the performance of naphtha cracking furnace. Chemical Engineering Research and Design, 94, 2015, 307-316.

[13] Robinson, P.R. Dolbear, G.E. Hydrotreating and hydrocracking: fundamentals, Practical Advances in Petroleum Processing, Springer, New York, NY, 2006, 177-218.

[14] Ghashghaee, M. Shirvani, S. Two-step thermal cracking of an extra-heavy fuel oil: experimental evaluation, characterization, and kinetics. Industrial & Engineering Chemistry Research, 57, 2018, 7421-7430.

[15] Barrie, J. Mullinger, P. Industrial and Process Furnaces: Principles, Design and Operation, Elsevier, 2011.

[16] Cambron, Adrien, and Colin H. Bayley. "Pyrolysis of the Lower Paraffins: II. The Production of Olefines in Baffled Quartz Tubes, Canadian Journal of Research 9.2 (1933): 175-196.

[17] Cambron, A. Bayley, C.H. Pyrolysis of the Lower Paraffins: III. Production of Olefines in Baffled Metal Tubes, Canadian Journal of Research 9, 1933, 583-590.

[18] Du, R.-L. et al. A modified Arrhenius equation to predict the reaction rate constant of Anyuan pulverized-coal pyrolysis at different heating rates, Fuel Processing Technology 148, 2016, 295-301.

[19] Gray, M.R. and McCaffrey, W.C. Role of chain reactions and olefin formation in cracking, hydroconversion, and coking of petroleum and bitumen fractions, Energy & Fuels 16, 2002, 756-766.

[20] Kuritsyn, V. A. et al. Modeling of pyrolysis of straight-run naphtha in a large-capacity type SRT-VI furnace,

Chemistry and Technology of Fuels and Oils, 44, 2008, 180-189.

[21] Savage, Ph.E. Mechanisms and kinetics models for hydrocarbon pyrolysis, Journal of Analytical and Applied Pyrolysis, 54, 2000, 109-126.

[22] Houser, T.J. Rate equations for reactions at high temperatures involving radical intermediates: consecutive-parallel and Rice–Herzfeld mechanisms, Journal of Chemical Physics, 50, 1969, 3962-3965.

[23] Baukal, Ch.E. Vaccari, M. Claxton, M.G. Burners for reformers and cracking furnaces, Computer Aided Chemical Engineering, 45, 2019, 937-984.

[24] Herman Francis, M. Mc Ketta, J.J. eds. Kirk-othmer encyclopedia of chemical technology, 1963.

[25] Habibi, A. Merci, B. Geraldine, J.H. Impact of radiation models in CFD simulations of steam cracking furnaces, Computers & Chemical Engineering, 31, 2007, 1389-1406.

[26] Oprins, A.J.M. Geraldine, J.H. Calculation of three-dimensional flow and pressure fields in cracking furnaces, Chemical Engineering Science, 58, 2003, 4883-4893.

[27] Amghizar, I. et al. New trends in olefin production, Engineering, 3, 2017, 171-178.

[28] Stöcker, M. Methanol-to-hydrocarbons: catalytic materials and their behavior, Microporous and Mesoporous Materials, 29, 1999, 3-48.

[29] Dry, M.E. The fischer–tropsch process: 1950–2000, Catalysis Today, 71, 2002, 227-241.

[30] Stanislav, J. Arellano-Garcia, H. Wozny, G. Oxidative coupling of methane in a fluidized bed reactor: Influence of feeding policy, hydrodynamics, and reactor geometry, Chemical Engineering Journal, 171, 2011, 255-271.

[31] Donazzi, A. et al. Microkinetic modeling of spatially resolved autothermal CH_4 catalytic partial oxidation experiments over Rh-coated foams. Journal of Catalysis, 275, 2010, 270-279.

[32] Zacharopoulou, V. Lemonidou, A.A. Olefins from biomass intermediates: A review, Catalysts, 8, 2018, 2-19.

[33] Al-Salem, S.M. Lettieri, P. Baeyens, J. Recycling and recovery routes of plastic solid waste (PSW): A review, Waste Management, 29, 2009, 2625-2643.

[34] Mohsenzadeh, A. Zamani, A. Taherzadeh, M.J. Bioethylene production from ethanol: A review and techno-economical evaluation, Chem BioEng Reviews, 4, 2017, 75-91.

[35] Chang, C.D. Methanol conversion to light olefins, Catalysis Reviews Science and Engineering, 26, 1984, 323-345.

[36] Haro, P. et al. Bio-syngas to gasoline and olefins via DME–a comprehensive techno-economic assessment, Applied Energy, 108, 2013, 54-65.